Pr

Finding

"Ken Miller is both an outst ... acher. In this scholarly book, he us ... ain the overwhelming evidence for evolution in a lucid and very entertaining way. In this process, he directly addresses the arguments that are made by each of the various schools of modern creationism. Dr. Miller is also a deeply religious person, and in a very personal and direct way he takes issue with those scientists who claim that modern science has disproven the existence of God. [Ken Miller] convincingly argues that science and religion offer different, but compatible, ways of viewing the world. In taking this position, he is supported by the leaders of most of the world's major religions. His book should be read by all those who want to understand this central issue."

—Bruce Alberts, president of the National Academy of Sciences

"Miller argues that evolution is the key to understanding our relationship with God, its indeterminacy endowing us with our freedom to choose good from evil and to choose God's love. Along the way, he pillories those religionists who see evolution as the product of iterative intelligent design and those atheists who argue that evolution demonstrates a universe lacking in purpose. An original, affective treatise, beautifully written, that offers new paths toward reconciling science with the faith of the Abrahamic tradition."

—Ursula Goodenough, professor of biology, Washington University, St. Louis; author of *The Sacred Depths of Nature*

"Religion's answer to Stephen Jay Gould's scientific atheism, Kenneth R. Miller, Brown's superstar in biology and religion, here shows not only why Darwinian evolution does not preclude the existence of God, but how remarkably consistent evolution is with religion. Written with sharp wit and in pungent prose, his book redefines the entire debate by showing the true meaning of the science represented by the name of Darwin. Had William Jennings Bryan read Miller's book, he would have not botched the Scopes trial—but then, there'd not have been such a trail to begin with."

—Jacob Neusner, Distinguished Research Professor of Religious Studies, University of South Florida and Bard College

"Can evolution and God coexist? With powerful logic and evidence, Kenneth Miller, a distinguished biologist and believer, develops an affirmative answer. *Finding Darwin's God* is an artfully constructed argument against both those who deny evolution and those using science to justify a materialist worldview. Yet it is a book for all readers. I know of no other that would surpass it in being mindful of different views, while still forceful. Miller has an uncanny gift for expressing profound ideas in clear and graceful prose."

—Francisco J. Ayala, Donald Bren Professor of Biological Sciences, University of California, Irvine

"*Finding Darwin's God* [is] a remarkable book that censures both fundamentalist creationism and the dismissive atheism of some scientists."

—*Philadelphia Inquirer*

"Part of the canon of evolutionary war literature."

—*Hartford Courant*

"An act of intellectual daring and spiritual integrity. . . . A refreshing departure from the tired polemics of the evolution wars."

—*Booklist* (starred review)

"Compelling, presented in terms that any layman could understand . . . never condescending or dull. . . . Convincing."

—*Christian Science Monitor*

FINDING DARWIN'S GOD

A Scientist's Search for Common Ground Between God and Evolution

KENNETH R. MILLER

NEW YORK • LONDON • TORONTO • SYDNEY

HARPER PERENNIAL

A hardcover edition of this book was published in 1999 by Cliff Street Books, an imprint of HarperCollins Publishers.

P.S.™ is a trademark of HarperCollins Publishers.

First paperback edition published 2000.
First Harper Perennial edition published 2002.
Reissued in Harper Perennial 2007.

Designed by Elliott Beard

The Library of Congress has catalogued the hardcover edition as follows:

Miller, Kenneth R. (Kenneth Raymond).
Finding Darwin's God: a scientist's search for common ground between God and evolution / Kenneth R. Miller.
p. cm.
Includes index.
ISBN 0-06-017593-1
1. Evolution—Religious aspects—Christianity. I. Title.
BT712M55 1999
231.7'652—dc21 99-16754

ISBN: 978-0-06-123350-0 (pbk.)
ISBN-10: 0-06-123350-1 (pbk.)

09 10 11 ❖/RRD 10 9 8 7

In the beginning was the mounting fire
That set alight the weathers from a spark

DYLAN THOMAS,
"In the Beginning"

CONTENTS

PREFACE

Introductions to Darwin's great work *On the Origin of Species* routinely tell readers that it belongs to a small group of books that have changed the face of the earth. As the distinguished biologist Ernst Mayr wrote for a 1974 reprint of the *Origin's* first edition, "Every modern discussion of man's future, the population explosion, the struggle for existence, the purpose of man and the universe, and man's place in nature rests on Darwin."[1] Evolution remains the focal point, the organizing principle, the logical center of every discipline in biology today.

Yet evolution also remains a point of concern and controversy, because it deals with the greatest of all mysteries, our own origins, and our human place in nature. The institutions of religion had once claimed solutions to these mysteries as their own, and the notion that natural science might find its own answers to such questions stirred immediate conflict. Darwin felt the conflict clearly, and attached three quotations to serve as epigraphs to the later editions of *Origin*. Each tells us something about Darwin's view of the proper relationship between religion and natural science, but the third, from Francis Bacon's *Advancement of Learning*, is particularly revealing:

[1] Darwin, C. D., *On the Origin of Species*, Fascimile of the first edition, with an introduction by Ernst Mayr. Cambridge, MA: Harvard University Press, 1974.

> To conclude, therefore, let no man out of a weak conceit of sobriety, or an ill-applied moderation, think or maintain, that a man can search too far or be too well-studied in the book of God's word, or in the book of God's works; divinity or philosophy; but rather let men endeavor an endless progress or proficiency in both.

In no small way, my purpose in writing this book has been to argue that Darwin chose exceptionally well when he selected this quotation.

The common assumption, widely shared in academic and intellectual circles, is that Darwinism is a fatal poison to traditional religious belief. One may, of course, accept the scientific validity of evolution and profess belief in a supreme being, but not without diluting traditional religion almost beyond recognition, or so the thinking goes. Incredibly, all too many traditional believers accept this view, not realizing that it is based more on a humanistic culture of disbelief than on any finding of evolutionary science. In a curious way, this allows each side to validate the extremes of the other. Religion leads one side to reject the cornerstone of the life sciences, while the other delights in telling us that science can determine the meaning of life—which is, of course, that it does not have one.

Lost in the fury is the hope Darwin expressed in his epigraph—that true knowledge could be found as much in "the book of God's works" as in the book of his word. I have written this book to make exactly that point, and leave it to the reader to judge if I have made it well.

Like any person who has been lucky enough to have the chance to pursue a scientific career, I am especially indebted to the many remarkable people who opened the doors and showed me the way. They include gifted teachers and patient mentors such as Richard Ellis, Andrew Holowinsky, Richard Goss, and Elizabeth Leduc. I am grateful to Andrew Staehelin and Daniel Branton, who guided me through scientific apprenticeships, and especially to the late Keith Porter, who expertly introduced me to the art and craft of science teaching.

During the preparation of this manuscript, my colleagues at Brown University were exceptionally generous with their advice, criticism, and most notably with their time. Included in this wonderful group are David Rand, Kristi Wharton, Marjorie Thompson, Susan Gerbi, Peter Gromet, Christine Janis, Walter Quevedo, and Alan Flam. I owe a special debt of thanks to a remarkable scientist, the late Jack Sepkoski, whom I came to know only over the past few years. Jack's patient studies of the fossil record set the standard for a new level of professionalism in paleontology, and his kind encouragement of my work on this book meant more to me than he could have known.

I am indebted to Robert Bloodgood for his thoughtful and careful reading of several sections of the manuscript, and also to my longstanding friend and colleague, Ursula Goodenough, a kindred spirit in the field of cell biology. I was also fortunate to have the kind assistance of Russell Doolittle to explain to me the intricacies of blood clotting, although I must assure the reader that any technical errors in my descriptions of these pathways are mine and mine alone.

A number of people were kind enough to supply material, figures, quotations, and ideas for this book from their own work, including Stephen Jay Gould, Hezy Shoshani, Phillip Johnson, Thomas M. Cronin, G. Brent Dalrymple, Harvey Lodish, and J. G. M. Thewissen.

I thank Barney Karpfinger, who believed in this project, nurtured it through my first halting attempts to plan a book, and helped to see it to completion. I appreciate the patience and kindness of Joseph Levine, my coauthor on many textbooks, whose support and friendship have been constants over the years. I am grateful to my editor, Diane Reverand, to my secretary, Jennifer Turner, and to all of my students over the years who have helped to keep my eyes firmly focused on the best ideals of science and learning.

Finally, I must thank the people who are and will always be close to me, including Marion and Ray, Lauren and Tracy, and most especially Jody, whose love, understanding, and support make all things possible.

I

DARWIN'S APPLE

Where are you from?" It's the kind of question that strangers, trying to become friends, will often ask one another.

No one can begin to know another until he knows where that person is from. Not just his family, school, and town, but everything that has helped to bring him to this point in his life.

This book is about the ultimate "Where are you from?" question. As important as it may be to understand one's ethnic origin and cultural identity, the bigger question is one that every child, sooner or later, asks of his or her parents: "Where did people come from?" In each culture according to its fashion, every child gets an answer. For me, a little boy growing up in suburban New Jersey in the 1950s, the answer came in the form of the first couplet of my religious training:

Question: "Who made us?"

Answer: "God made us."

Every year, that training reached deeper, demanded more, and grappled with more sophisticated questions of faith and virtue. But every year, it began with exactly the same question: Who made us? And that question was always followed by exactly the same answer. *God made us.*

In a different building, only a few hundred yards away from the red brick walls of St. Mary's, I began to find another answer to that question. This other school did not always grapple with the same straightforward questions of right and wrong that were the weekly fare of our catechism, but it taught its students to believe something at least as intoxicating as the divinity of their origins—the possibility that the world around us was constructed in such a way that we could actually make sense of it. That great secular faith drew strength from a culture in which science seemed to fuel not only the fires of imagination, but the fires of industry as well. And that faith extended to living things, which yielded, like everything else in the natural world, to the analysis of science.

Looking back on my youth, I am struck by how meticulously those two aspects of education were channeled to avoid conflict. Teachers on both sides, secular and religious, were careful to avoid pointing out the dramatic clash between the most fundamental aspects of their world views. No one ever suggested a catechism with a different beginning:

Question: "Who made us?"

Answer: "Evolution made us."

Nonetheless, the conflict between those two points of view is real. The traditional Western view of humanity as the children of God once had a direct, literal basis in the historical narrative of sacred scripture. Not only was God our spiritual father, He was also the direct agent of our creation. His actions were the immediate cause of our existence, and His planning and engineering skills were manifest in every aspect of our bodies. By extension, the splendor and diversity of the living world that surrounds us testified to the very same care and skill.

Charles Darwin himself recognized how profoundly scientific analysis had changed this view of life and humanity when he wrote the historical sketch that preceded his great work, *The Origin of Species by Means of Natural Selection.* Generously (and correctly) he gave credit for this transformation to the now much-maligned French naturalist Jean Baptiste Lamarck:

> In these works, he [Lamarck] upholds the doctrine that all species, including man, are descended from other species. He first did the eminent service of arousing attention to the probability of all change in the organic, as well as in the inorganic world, being the result of law, and not of miraculous interposition.[1]

Today it is very clear that the line of reasoning Darwin attributed to Lamarck has emerged triumphant. Change in the inorganic and organic world is no longer attributed to "miraculous interposition." It once was possible to point to a humble seed and invoke the attention of the Almighty as the only possible explanation for how such an ordinary object could grow into a mighty tree. Today we look into the seed itself, examine the program of gene expression that begins at germination, and seek our answers in the rich complexities of molecular biology and biochemistry. This does not mean that we have reduced the seedling to mere chemistry or physics. It means instead that we have elevated our understanding to appreciate the living plant in a way that lends wonder and delight to our view of nature.

My purpose in this book is to attempt something that is generally avoided. I want to ask a question that most of my colleagues shy away from, and to attack head-on the defenses that many of us have built up in our unwillingness to reconcile the two different answers to the question of "Who made us?" The question is whether or not God and evolution can coexist.

There is no need to break new scientific ground in approaching this question. The century and a half since Darwin has provided us with more than enough time to flesh out the details of his abstract outline on the process of biological change. To add to Darwin's ideas we have half a century of molecular biology, bold explorations of space and time provided by the physical sciences, an understanding of earth's history from geology, and even an appreciation of the limits of our most powerful reasoning tool—mathematics. We have to be willing to bring all these resources to bear in unfamiliar surroundings, to apply them in new ways, and to ask the sorts of questions that are not commonly heard in scientific circles.

We can by starting with the man himself, Charles Darwin, a writer of exceptional clarity whose words and ideas remain accessible, even today.

Summer Reading

My first encounter with Charles Darwin cost me a buck ninety-five, the price of a Penguin paperback copy of *On the Origin of Species*. I picked the book up one July day in the summer of 1966 in a New Jersey mall. Suspended in that strange summer between high school and college, I had found just the place to spend twelve pleasant weeks converting my one marketable skill, lifeguarding, into enough spare cash to get through my first year at university. When the chance presented itself, I supplemented my not-exactly-generous lifeguard's pay by giving private swimming lessons to the sons and daughters of club members.

Those swimming lessons helped to put enough extra change in my pocket to afford to buy a book like Darwin's. Reading "serious" books had become one of my most important projects that summer. For twenty minutes each hour, one of the three lifeguards on duty was rotated into a large booth at the club entrance where we were to check IDs and "greet club members with enthusiasm." After the morning rush on sunny days, there was plenty of time to socialize with kids my own age or just lean back and read.

Visible reading was part of my other important "project," a young lady at the swim club. Early in our mutual flirtation, she made it clear to me that she regarded reading serious books as a mark of "depth." From that moment on, in the parlance of the time, I wanted to be as "deep" as possible—at least while she was looking. My reading list that year included not just Darwin, but Augustine, Eliot, Marx, Durant, Shirer, Milton, and Dante. I must have been quite a sight, propped up in the corner of the check-in booth stumbling through the likes of *The Wasteland,* always making sure that the title was in plain view just in case my young love strolled past unexpectedly.

I remember that summer with something between amusement and embarrassment. I managed to stumble through those books, collect a tan on the lifeguard perches, survive love's heartbreak, and as the days began to shorten, pack myself off to my first year of college.

However impure my motives for tackling those volumes, something did sink in. Several books took hold so thoroughly that I wound up finishing them at home, sitting on our backyard porch listening to Mel Allen's fluid descriptions of New York Yankee night games in the background. Each book had its own life, and each carved out its own niche in my imagination. Shirer's descriptions of prewar Berlin enlivened my father's war stories; Durant convinced me that philosophy was a lively subject, despite the density of its language and thickness of its books; and Saint Augustine made me realize that even in the fifth century A.D. there lived people who had the same likes and dislikes, the same difficulties with authority, and the same weaknesses that I had. My personal favorite, unexpectedly, was Milton, whose *Paradise Lost* at first seemed to drone through line after line of pointless extrapolation exploding from the clarity of the blessedly concise Book of Genesis. Then, just as I was about to put it down, I happened upon Milton's wonderful passage about God's decision to create Eve.

Surrounding Adam in the Garden were the other creatures of God's creation, the lion and the lamb, the snake and the hummingbird, all wrapped in the harmony of Eden. As Adam was quick to note, there was something special about him. Not his intellect, not his cleverness, not his likeness so close to God. No, the difference that came to the mind of Milton's Adam was his loneliness. He alone, among the creatures of Eden, was the solitary member of his kind. Boldly, Adam brought his complaint to the Almighty:

Let not my words offend thee, Heav'nly Power,
My Maker, be propitious while I speak.
Hast thou not made me here thy substitute,
And these inferiour farr beneath me set?
Among unequals what societie
Can sort, what harmonie or true delight?[2]

What society, indeed? Adam was quick to realize that Eden, however glorious, included little for him in the way of conversational possibility. Wandering alone in God's fair Garden might be a fine way to pass the time, but Adam, as it turned out, had a little more on his mind:

Of fellowship I speak
Such as I seek, fit to participate
All rational delight, wherein the brute
Cannot be human consort; they rejoyce
Each with thir kinde, Lion with Lioness;
So fitly them in pairs thou hast combin'd . . .[3]

To my amazement, Milton's God acceded to Adam's request, and agreed to heal his loneliness with a new creation—Eve. Things then got complicated, we might say, in ways that most of us know all too well. John Milton described the mother of mankind in terms quite different from the modest prose of Genesis:

Shee as a vail down to the slender waste
Her unadorned golden tresses wore

Dissheveld, but in wanton ringlets wav'd
As the Vine curles her tendrils, which impli'd
Subjection, but requir'd with gentle sway,
And by her yielded, by him best receivd,
Yielded with coy submission, modest pride,
And sweet reluctant amorous delay.
Nor those mysterious parts were then conceald,
Then was not guiltie shame, dishonest shame
Of natures works, honor dishonorable . . .[4]

To an eighteen-year-old whose personal discoveries that summer were taking a pathway in the imagination remarkably similar to those of the fictional Adam, John Milton's verse had all but burst into flame. I began to think that packed in the midst of this epic work, so respected and admired by a dusty educational establishment, there just might be the forbidden fruit of desire, passion, and that most unscholarly of all qualities—humor—stunning, memorable, ironic. Maybe, just maybe, if you read through these great, gray books you eventually got to the good parts.

MISSING THE POINT

The only scientific book I read that summer was Charles Darwin's *The Origin of Species*. Because I am a biologist, one might expect me to say that the power and subtlety of this great work drew me into the life of an experimental scientist. Nothing could be further from the truth. I found Darwin *boring*. You might say that I kept looking for the good part, something that would match those sensual passages from Milton, and I never found it. At least not that summer.

It was clear to me after the first few pages that the book by this nineteenth century naturalist was not in the same rhetorical league as the rest of my poolside syllabus. In its own way, that made perfect sense. What book of science was? Ordinary people, after all, no longer read Newton's *Principia* or Lyell's *Principles of Geology,* even though we continue to delight in the craft of their literary contemporaries, Swift and Dickens. But we do still read Darwin. That must mean, to draw the obvious conclusion, there was something important in the *ideas* between the covers of this classic.

I remember reading the first three chapters of *The Origin* page by page before I lost patience with the author. Mercifully, I discovered the careful summaries tucked away at the end of each chapter. I charged to the end of each chapter of the work, where every student's dream, a summary of the summaries, awaited me. Dutifully I worked through them, and finally, one rainy afternoon, surveying an empty pool and an abandoned parking lot, I decided that I was finished with Charles Darwin. Everything in *The Origin of Species* had seemed just too ordinary.

Nothing in my adolescent reading of *The Origin* could match the sensual poetry of Milton, the brooding darkness of Eliot, or the chilling spell of Dante's admonition above the gates of hell: "Abandon all hope, O ye who enter here!" Pitted against these masterpieces, all that Darwin had to offer was a common-sense rendition of one obvious observation after another on the nature of living things.

The man's arguments—and he called his book "one long argument," went something like this:

Domesticated plants and animals show a tremendous range of variation. This was obviously true. After all, variation is the raw material upon which the breeders of animals and plants are able to work. By selecting, consciously or unconsciously, the individuals who will give rise to the next generation, they gradually form new and distinct varieties, which can differ so greatly from one another that they are barely recognizable as members of the same species.

Take, for example, two common breeds of dog: the Great Dane and the Chihuahua. There isn't the slightest doubt that they are both dogs (*Canis familiaris*) or that they are both descended from common ancestors. Yet think how a naturalist who had never seen a dog would respond to these two creatures! Without a doubt, he would quickly conclude that they were different species. That's variation!

A similar range of variation exists in nature among wild species. I wondered about this, and was ready to "challenge" Darwin, but he summed it up in a convincing way—by pointing out that the variation was so great that naturalists argued endlessly among themselves as to whether the individuals of a widely dispersed type were one species or two.

> Compare the several floras of Great Britain, of France or of the United States, drawn up by different botanists, and see what a surprising number of forms have been ranked by one botanist as good species, and by another as mere varieties.

At that point in my life, I had had only the slightest glimpse of academic science, but even this was enough to support Darwin's view. Arguments over exactly how different two populations had to be to constitute separate species were common. As it turned out, these natural variations were more than just a little important to Darwin. To make sure that we would not mistake what he was driving at, in Chapter 2, titled "Variation in Nature," he wrote:

> These differences blend into each other in an insensible series; and a series impresses the mind with the idea of an actual passage.[5]

An *actual passage*. How about that? But let's not get ahead of ourselves.

All living things are engaged in a struggle for existence. Like anyone who has ever tended a small garden, I immediately knew that Darwin was right about this. I might have been only eighteen, but I had already watched tomatoes die under an onslaught of cutworms, red ants dismember a hapless beetle under my magnifying lens, and scores of weeds and grass seedlings sprout only to wither and vanish under the sharp August sunlight. Despite all this carnage, our small yard teemed with life.

Darwin explained it all succinctly. First, living things can produce more of themselves.

> There is no exception to the rule that every organic being naturally increases at so high a rate, that, if not destroyed, the earth would soon be covered by the progeny of a single pair.[6]

Because only a few of those progeny can survive, there is a struggle for existence among them—the same struggle I had seen daily in the garden. And finally, a key insight: The struggle is most severe among individuals of the same species. Why? Because the members of your own species are the very ones who need *exactly* the same resources you do to survive. In short, to know what your number one competitor is like, take a look in the mirror—he's going to look a lot like you.

This struggle, combined with variation, results in natural selection. Darwin began the fourth chapter of *The Origin,* "Natural Selection," with a rhetorical question: "How will the struggle for existence, briefly discussed in the last chapter, act in regard to variation?"[7]

Darwin's answer was that it would act automatically. Those individuals that lose in the struggle for existence generally do not get to produce the next generation, but those individuals that do succeed get the greatest of all possible rewards—they get to pass their winning traits along to their offspring.

This means that the conditions of nature, whether acting in my backyard, on the Galápagos Islands, or atop Mount Fuji, are constantly acting on natural variation, selecting out unsuccessful variations and rewarding successful ones. When forces divide a single species into two populations, natural selection will act on each separately, until they have accumulated enough differences that each becomes a separate (and new) species.

Incredibly, that's all there was to it. In those principles you have all of Darwin's theory. Being a long-winded Englishman, Darwin wasn't going to end with just four chapters and a hundred pages. He went on for eleven more chapters, explaining in numbing detail the implications of his theory for biogeography, paleontology, classification, instinctive behavior, and embryology. He even considered objections to his theory, and, just in case you had missed something important, fashioned a concluding chapter to recapitulate his arguments.

I did have to admit that each of the four building blocks of this theory was obviously true. Breeders did draw upon the range of variation in domestic animals and plants to make new varieties. A similar range of variation did exist in nature. The conditions of life did place each individual in competition with others. And this competition clearly affected the range of variation that survived.

Thomas Henry Huxley, soon to be called "Darwin's Bulldog" for his determined advocacy of Darwinism, is said to have read *The Origin* and then to have remarked, "How foolish of me not to have thought of it!" I would love to say I felt the same way, but I can do nothing of the sort. Darwin's observations seemed tediously obvious. They also led to an obvious conclusion—that all life was interrelated, which hardly seemed earth-shattering to me at the time.

American scientific education, particularly during my youth in the fifties and sixties, is often assumed to have been so timid on the issue of evolution that a whole generation grew up knowing next to nothing about the subject. There may have been some truth to that, but like most kids my age, I still had a pretty good understanding of earth history. That understanding had been reinforced by a handful of visits to fossil collections at the American Museum of Natural History in New York. The bones and shells and especially the great dinosaur reconstructions left very little doubt that life on earth had once been very different from today. Before long we all were convinced that evolution was the process that had produced the dramatic changes the museum documented in such spectacular fashion.

My encounter with Darwin's detailed, careful, exhausting nineteenth century prose was less than memorable the first time around. At best, it gave me a sense of where these rich and wonderful ideas of our past had come from. But like most of my contemporaries at the time, I placed little value on historical context. I found nothing in *The Origin* worth reading to my love, which, for me, made it the single most boring book of an otherwise memorable summer.

A Dangerous Mind

In retrospect, there was just one thing that made me reluctant to put down Darwin's book, something extraordinary that kept me going until I had read at least a bit of each chapter, all of the conclusion, and then revisited each of the summaries. This motivating force was something I had not encountered in all the rest of my reading. People were *afraid* of the book.

My dad, whose chance at college came only when the end of World War II allowed him to lay down a rifle and pick up the GI bill, had ignored most of my attempts to act like an intellectual that summer. When he saw me reading Darwin, he thoughtfully told me that I was reading a dangerous book, and I should be careful. I wondered about that. A few people at the swim club made the same comment. "Be sure you talk to someone about that book," a member solemnly warned, "and be careful not to lose your values." These warnings made the book seem more interesting at first, and kept me pumping through the first few chapters searching for scandal. I never found the good parts that would have justified such concerns. The scandal, no matter how carefully I read, escaped my notice.

This was quite a disappointment, but my father's warning did have a certain ring of truth to it. I had taken two public high school courses in biology. My first biology teacher, Paul Zong, was in love with life, and wanted to draw us into his kind of biology—a fellowship of systematic exploration, classification, and nomenclature. I am positive, even after I have earned two university degrees and accumulated years of specialized training, that my knowledge of classification and scientific names reached its high-water mark on the morning of my final exam in ninth grade biology. By the time I had ended Paul Zong's class, I knew exactly what I wanted to be—a biologist. And I never changed my mind.

I took my second course in biology, an advanced course, as a senior from an uninspiring instructor whose name is best not mentioned. This awful class would pass unmentioned except for one striking similarity it shared with my wonderful year under Mr. Zong. Neither teacher mentioned evolution.

Many years later, wondering if that recollection had been correct, I retrieved the exact edition of the biology textbook we used for the first of these courses. The word "evolution" did not appear in the index. Charles Darwin was mentioned, but only in a chapter strangely titled "Organic Variation through Time." The facts of natural history were too compelling to skip completely, but the mention of evolution was carefully avoided. This was essentially the strategy followed by both of my teachers. I did indeed learn about natural selection, the human fossil record, and the age of the dinosaurs. I simply had not learned *evolution* as a specific subject.

This was hardly a unique experience. Historians of American scientific education would later document the great retreat from evolution that swept over public education at midcentury. I suspect that at one time or another in their careers, my teachers had discovered that teaching evolution and calling it such brought concern and discomfort to an otherwise quiet and respectable profession. Like most of us, they found an easier way through the woods, and then trod that path year after year.

The dearth of apparent scandal in *The Origin* made all of this a little puzzling, and it took me a few years to understand what all the fuss was about.

During my second year of graduate school at the University of Colorado, trying to make ends meet, I was teaching a regular tutorial session of freshman biology students in the dorm lounge several nights a week.

One day, two of my students wandered into my advisor's lab holding a tiny newsprint pamphlet. I have long since forgotten the specifics it contained, but the cover illustration is plain as day in my recollection. "Evolution—The LIE" was emblazoned on the surface of an apple, while a snake coiled about it with obvious menace.

Not only was evolution linked with the Father of Lies himself, it was "exposed" as a massive conspiracy foisted on an unsuspecting public. In case I had been thick enough to wonder how a specialized biological theory might play such a crucial role in society, the back cover of the pamphlet spelled it all out for me. Evolution, it seemed, was responsible for such evils as theft, murder, drug abuse, prostitution, war, and even adultery. I dismissed the pamphlet with a sneer, cleverly pointing out to my younger friends that the Old Testament documented nearly all of these sins long before *The Origin,* and it seemed a bit much to blame Darwin for all of them.

The pamphlet jarred a nerve within me, however, put there no doubt by years of careful study of the catechism. The dangers sensed by the pamphleteers, by my Dad, and by my teachers all had the same source. The danger in evolution was that it struck directly at the fundamental assumptions of religion about the relationship between God and man. Evolution threatened the soul itself.

The Apple Drops

Darwin's "dangerous idea," as philosopher Daniel Dennett called it in his recent book by that name, is surely one of the most influential and far-reaching ideas of all time.

Although Darwin was careful to note the contributions of those who preceded him, the publication of *The Origin* was a public event unprecedented in the history of science. He may have hoped to write a book to be read and appreciated by specialists, but it immediately became a widely discussed best-seller. Evolution turned out to be hot stuff. Its influence has been felt in fields as disparate as immunology and sociology, and it has revolutionized the way in which we view the world, natural and man-made. Dennett accurately described the impact of Darwin's theories:

> Let me lay my cards on the table. If I were to give an award for the single best idea anyone has ever had, I'd give it to Darwin, ahead of Newton

> and Einstein and everyone else. In a single stroke, the idea of evolution by natural selection unifies the realm of life, meaning, and purpose with the realm of space and time, cause and effect, mechanism and physical law. But it is not just a wonderful scientific idea. It is a dangerous idea.[8]

We'll return to visit some of Dennett's ideas on the "danger" represented by the theory of evolution in Chapter 6, but here we will consider some of the undeniable power and scientific fruitfulness that evolutionary thought offers.

One of the great beauties of evolution is that it is automatic. The combination of random variation and natural selection automatically selects the organisms that do best in a particular environment, and then rewards them by forming the next generation from the winners in the game of natural selection. The power of this simple idea extends well beyond biology.

When I first began university teaching, I was unexpectedly impressed by the skill of my students at taking tests. Test-taking is an ability distinct from actual knowledge, which is not to say that knowledge is unimportant to success on a well-constructed exam. There is a strategic component to doing well, and all of my students seemed to excel at this. They asked the right questions: "Is there a correction [on the multiple choice questions] for guessing?" "Should I write a literal answer, or do you want us to extrapolate?" "Do you want all possible solutions? Or just the most obvious one?" They paced themselves intelligently, quickly identifying the most time-consuming questions, making sure that they left them for last, scoring as many points as they could on questions that could be quickly answered.

Did we have a course where they are taught how to take tests, I wondered. I asked this out loud to a colleague, who laughed. "Don't have to. It's natural selection," he said with a grin.

Why natural selection? Student admissions at the university where I teach are extremely competitive. Our students are products of an educational system in which those who are proficient at test-taking are moved towards the honors tracks in middle school and high school. Once there, they can earn the best grades only by excellent exam work—test-taking again. If that were not enough, nearly all of our students, number two pencils in hand, take a nationwide aptitude test as the capstone of their preparation. Students who are not proficient at test-taking are gradually weeded out. By the time they reach my classroom, only the adept test-takers are left. It works automatically, just like natural selection.

Biologist and author Richard Dawkins once allowed his readers to consider how powerful natural selection was by asking them how many of their direct ancestors had died in childhood.[9] When I pose this question to my own students, they take it seriously at first, and I can almost see the mental wheels spinning as they silently reconsider the lives of their parents and grandparents. Within a few seconds, most of them are grinning or laughing out loud as they realize how ridiculous such a question is. *None* of anyone's immediate ancestors died in childhood. If they had, they wouldn't be your ancestors!

Putting it another way, each of us is descended from a long line of winners. The near-perfection our bodies display as they grow, processing food, regulating temperature, and resisting disease, has been honed at a price. Unsuccessful experiments in metabolism or design, like poor test-takers, are weeded out by natural selection.

Darwin's powerful idea, a biological explanation for the origins of living species, has exerted a transfixing hold on human thinking in the century and a half since it first kicked in the doors of Western intellectual life. Once Darwin's apple had fallen from the tree, there was no stopping the ways in which eager scholars would apply it to one problem after another. Like a tide sweeping away old explanations of natural philosophy, Darwinian thought made scientists everywhere demand naturalistic, materialist explanations for the way things are. The intellectual dominance of these ideas led to a new set of cultural assumptions about science, about the world, and even about the nature of reality.

Was this because Darwin provided the first workable explanation for the remarkable adaptations of living things? I suspect not. I also do not believe that Darwin's wide influence comes from his patient and groundbreaking observations on orchids or barnacles. Rather, it comes from one simple fact. Evolution displaced the Creator from His central position as the primary explanation for every aspect of the living world. In so doing, Darwin lent intellectual aid and comfort to anti-religionists everywhere. As Dawkins accurately observed:

> Although atheism might have been *logically* tenable before Darwin, Darwin made it possible to be an intellectually fulfilled atheist.[10]

For nonbelievers like Richard Dawkins, Darwin provided the first complete, rational basis for rejecting the spiritual and the supernatural. In a Copernican universe where Newtonian laws, not the chariots of gods, moved the heavens around, evolution disposed of the last remaining mystery—the source of life itself. With that taken care of, surely there was no longer any room left for religion in the life of the mind. The world had at last been made safe for "intellectually fulfilled" atheists. That is what was dangerous about *The Origin*.

The First Church of Charles the Naturalist

My particular religious beliefs or yours not withstanding, it is a fact that in the scientific world of the late twentieth century, the displacement of God by Darwinian forces is almost complete. This view is not always articulated openly, perhaps for fear of offending the faithful, but the literature of science is not a good place to keep secrets. Scientific writing, especially on evolution, shows this displacement clearly.

In 1978, Edward O. Wilson won a Pulitzer Prize for his influential book, *On Human Nature*. Wilson's widely admired pioneering work on social behavior in insects led directly to his founding contributions to a new field which he called "sociobiology"—the study of the biological basis of social behavior. *On Human Nature* was a bold attempt to apply sociobiological principles to human behavior and human institutions. Right up front, on the very first page of the book, Wilson put his finger on the crux of the matter:

> If humankind evolved by Darwinian natural selection, genetic chance and environmental necessity, not God, made the species.[11]

For thousands of years, human beings thought of themselves as the children of God. After Darwin, they were the children of "genetic chance and environmental necessity." Wilson's sentiment that Darwinian natural selection rules out God is widely shared. Richard Dawkins leaves no doubt about his own view of a Darwinian universe. It is not a place of real values, of genuine good and evil. As Dawkins admits:

> This is one of the hardest lessons for humans to learn. We cannot admit that things might be neither good nor evil, neither cruel nor kind, but simply callous—indifferent to all suffering, lacking all purpose.[12]

The writings of scientists like Wilson and Dawkins present a viewpoint in which the need for God as an explanation for the nature of things is a relic of humanity's intellectual childhood, and nothing more.

A few years ago, an opinion column in *The Scientist,* a trade weekly for scientific professionals, had maintained that any scientist who entered a house of worship had better "check his brain on the way in." Although many readers of this publication stated views to the contrary, there was strong, well-argued support for the position that no educated scientific professional could possibly profess a belief in the supernatural. The readers of *The Scientist* are no fools—they knew exactly what had happened in recent scientific history. First Galileo and Copernicus displaced man as the center of the universe. Then Darwinism set aside God as the author of creation. And finally the rise of biochemistry and molecular biology removed any doubt as to whether or not the properties of living things, humanity included, could be explained in terms of the physics and chemistry of ordinary matter. The word is out—we are mere molecules.

So complete is the absence of the spiritual from modern science that a few writers even take withering potshots at God Himself. George C. Williams, a scientist who has made important contributions towards understanding the complexities of natural selection, did so in his book *The Pony Fish's Glow*:

> She [anthropologist Sarah Hrdy] studied a population of monkeys, Hanuman langurs, in Northern India. Their mating system is what biologists call harem polygyny: Dominant males have exclusive sexual access to a group of adult females, as long as they can keep other males away. Sooner or later, a stronger male usurps the harem and the defeated one must join the ranks of celibate outcasts. The new male shows his love for his new wives by trying to kill their unweaned infants. For each successful killing, a mother stops lactating and goes into estrous. . . . Deprived of her nursing baby, a female soon starts ovulating. She accepts the advances of her baby's murderer, and he becomes the father of her next child.
>
> Do you still think God is good?[13]

The terrifying infanticide practiced within the langur harem, as Williams explains, is a straightforward prediction of evolutionary theory. If it is genetically programmed, then it is an adaptive behavior because it increases the proportion of the new harem-master's genes in the next generation.[14]

A nonscientist reading the popular books of writers like these might be forgiven for jumping to the conclusion that modern evolutionary science has ruled out the existence of God. According to Williams, science *certainly* has ruled out the existence of a benign one.

Is this indeed the case? Is it time to replace existing religions with a scientifically responsible, attractively sentimental, ethically driven Darwinism—a First Church of Charles the Naturalist? Does evolution really nullify all world views that depend upon the spiritual? Does it demand logical agnosticism as the price of scientific consistency? And does it rigorously exclude belief in God?

These are the questions that I will explore in the pages that follow. My answer, in each and every case, is a resounding *no*. I do not say this, as you will see, because evolution is *wrong*. Far from it. The reason, as I hope to show, is because evolution is *right*.

2

EDEN'S CHILDREN

Not many years ago, astronomer Robert Jastrow cautiously began a book on the religious implications of scientific discovery. In its opening chapter he worried openly, "When a scientist writes about God, his colleagues assume he is either over the hill or going bonkers."[1] Jastrow knew that within the academy anyone, particularly a scientist, who openly reveals even the slightest twinge of religious conviction is regarded as a bit of a flake. Things haven't changed. I teach a large biology course at one of our country's great colleges, and every time I step to the podium I realize that those bright young faces in the audience are making assumptions about me. Sometimes, I hope, those assumptions are grounded in reality. I hope, for example, that they accept my enthusiasm for science as genuine, and that they detect the optimism in my voice when I talk about the future.

The teaching of evolution brings forward a new set of assumptions in my students. They very quickly come to believe that a professor who speaks with excitement and passion on evolutionary theory, who documents the rich fossil record of our prehuman ancestors, and who describes the role of chance and contingency in the origin of species probably is or must be an atheist. A few of them get the courage to stay after class and ask that personal question point-blank. And they get back an answer few of them seem to anticipate. "No, I'm not."

College students today pride themselves on their open-mindedness. Their intellectual instincts embrace the modern virtues of tolerance and diversity almost without question. Despite the conviction in some quarters that these graces extend only to one side of the spectrum of value and belief, I believe that most students are sincere in their attempts to be intellectually tolerant. When I answer their questions about evolution and religion by telling them that I happen to believe in God, I am treated respectfully and courteously. Nobody seems to get upset, no organizations picket my classes, my name does not appear on protest leaflets. Nonetheless, after explaining my personal beliefs at the request of a student, almost every time I get the response, "You believe that *despite evolution?*" Yup. *Despite* evolution . . . or maybe, as I will explain in the later chapters of this book, *because* of evolution.

Over years of teaching and research in science, I have come to realize that a presumption of atheism or agnosticism is universal in academic life. From time to time I have to struggle to explain to my students, and even my colleagues, not only why Darwinian evolution does not preclude the existence of God, but how remarkably consistent evolution is with religion, even with the most traditional of Western religions. It is simply taken for granted that smart, modern, well-informed people have risen above the level of petty superstition, which is exactly how any serious faith is regarded. Religion as *culture,* in the sense of Jewish culture, Islamic custom, and even Christian tradition, may be grudgingly accorded obligatory respect—just enough, I am sure, to evade the nasty charge of cultural imperialism. But religion itself, genuine belief, just doesn't belong.

It would be difficult to overstate how common this presumption of godlessness is, and the degree to which it affects any serious attempt to investigate the religious implications of ideas. Ironically, if asked to justify such attitudes, my friends and colleagues in nonscientific disciplines will often claim science as their authority. Clearly they believe that scientific inquiry has ruled out the divine. Unfortunately for them, as I will argue, nothing of the sort is true. Their attitudes towards religion and religious people are rooted not so much in science itself as in the humanist fabric of modern intellectual life. This is a point we will explore in detail a little bit later, and as we will see, it is the source of a powerful backlash that science ignores only at its peril.

Plain Talk

My apparent dissent from the dominant material certainty of evolution requires an explanation. Do I have some reason to doubt evolution? Is there some flaw, large or subtle, in Darwin's great theory, some new evidence that an embarrassed scientific community has hushed up? Many critics of evolution certainly seem to think so. In 1996, when writer David Berlinski attacked evolution across the board in *Commentary* magazine,[2] letters of praise from delighted readers poured in. They lauded his "courage," his willingness to confront a "stifling scientific establishment," and his achievement in breaking out of the "intellectual straitjacket" that Darwinism has imposed on modern life.

Berlinski is not alone. Such critics as Berkeley law professor Phillip Johnson have charged that the weakness of evolution's scientific standing is a trade secret known only in the inner circles of the establishment. An honest appraisal of the evidence, they maintain, would sweep evolution convincingly into the dustbin of history. Are these critics right? If there is indeed a major flaw in Darwin's edifice, why haven't scores of young, iconoclastic scientists, eager to establish their own careers, smashed it to pieces? These are questions worth answering.

For more than twenty years, I have been fortunate to make my home in New England. The town in which my family lives was founded only a couple of decades after the Pilgims landed, and it has remained just distant enough from nearby urban centers to retain its rural character. Although growth has begun to change the town, not always for the better, Rehoboth has recently suffered a slow plague of urban workers putting up homes on what used to be fields and woods worked for farming and lumber. Nonetheless, Rehoboth retains its essential New England character, much to the delight of its residents.

Although some of our region's educational institutions, my own included, have well-deserved reputations for political liberalism, New Englanders on a personal level are exceptionally conservative. New England leads the United States in church attendance and ice cream consumption, and in those seemingly unlinked characteristics I find a great deal to admire. The New England character has been formed by difficult weather and rough landscape. It is a place where one state has emblazoned "Live Free or Die" on its license plates,[3] and it is the only part of the United States where the beginning of the American Revolution is still a legal holiday.

My friends and neighbors accept all of this without question, and also take it as their right to control every aspect of our local government. If fact, they *are* the local government. The second Monday in April, a time when snowdrifts are unlikely to prevent travel but it's still too early to work the fields, is Town Meeting Day. In numbers that rise and fall with the level of local controversy, citizens gather in a school gymnasium to vote, line by line, on every item in the town budget. The police chief who needs a new radio, the tax collector who has her eye on a faster computer, and the building inspector whose truck has broken down must stand and plead their cases in front of the skeptical townfolk.

I've never had to go in front of the town meeting to talk about evolution, but if I did, I suspect that the questioning would be very practical. My plain-speaking neighbors would not much care about the philosophical implications of the idea. They wouldn't worry about the personal religious beliefs of Charles Darwin, and they wouldn't give a damn if other people had used evolution to support or justify their own views on religion, on economic systems, on linguistics, or anything else. Very much like my first-year students, the fundamental questions for my neighbors would be, sensibly, the big ones. The bottom-line issues. Did evolution actually take place? And if it did, did evolution produce us, too? In plain language, I would answer those questions the only way that fact and science allow. Yes, it did. And yes, we are the children of evolution, too.

Gumshoe Science

Can we be *absolutely* certain of this? In the strictest sense, *no*. Scientific knowledge, in the absolute sense, is always tentative. Science is founded on the proposition that *everything* we think we know about the natural world can, in principle, be rejected if it does not meet the test of observation and experiment. The very practice of science, at its core, is a constant exercise of extending what we do know about the world, and then correcting what we thought we knew for sure.

In this respect, evolution occupies no special place in the scientific hierarchy of ideas. Like any theory, including the germ theory of disease or the atomic theory of matter, it would have to be discarded and replaced by a better theory if scientific evidence made it clear that evolution was not an adequate explanation for the history of living organisms. There is nothing extraordinary in what I have just said. It simply means that evolution is a scientific idea, and scientific ideas rise or fall on the weight of the evidence.

Unfortunately, there is a school of thought that rejects the very idea that any theory about the past can be scientific. Science, the argument goes, is based on experiment and direct, testable observation. Therefore, science can address only phenomena that are brought into the laboratory and examined under controlled conditions. It is true that the very best experimental conditions are those in which we can control all variables, then manipulate just one or two, and make detailed observations of the results; but this argument would deny scientific inquiry to any situation that does not lend itself to laboratory science.

The natural history of the earth is just such a situation. Since there were no human witnesses to the earth's past, the argument goes, all statements about that past, including evolution, are *pure* speculation. Your speculation, some say, just like your taste in music, has no more claim to be scientific than my taste does. This line of reasoning, by denying scientific legitimacy to evolution, would take away from Darwinism its scientific cachet and place it in the same category as *any* story of origins. All answers to our "Where are you from" question would then have equal scientific standing—which is to say, no scientific standing at all.

As a practicing scientist, I might not like the implications of that argument, and I could oppose it just for that reason. I'd rather argue from a higher standard, however, which is whether or not the argument is correct. Is scientific inquiry restricted to what we can actually bring into the laboratory and see happening right in front of us? Is there really any scientific way that we can know *anything* about the past at all?

There is indeed a way to do this, and the process is so ordinary that most of us take it for granted. When a common burglar pries open a window and climbs into a home to steal, he does so in the hope that he will leave as he entered—silently and without detection. But despite his best efforts, every one of his actions leaves something behind. The marks of his entry will bear the imprint of his tools. His footsteps will leave impressions in carpeting, residue on wood, and telltale debris on both. The police may be able to gather samples of his hair, his fingerprints, perhaps even his blood. And sooner or later, the goods he took will reappear in a marketplace where they may be traced and identified as stolen.

A police detective would scoff at the notion that crimes can be solved only when they are witnessed directly. Not a single person witnessed Timothy McVeigh park a rental truck in front of the Murrah Federal Building in Oklahoma City on April 19, 1995. Nonetheless, when McVeigh was apprehended on an unrelated charge, investigators quickly assembled a web of evidence that left little doubt that he had been deeply involved in the planning and execution of the bombing. The simple fact is that we can learn about the past by applying good, old-fashioned detective work to the clues that have been left behind.

The same rules apply to science. We may not be able to witness the past directly, but we can reach out and analyze it for the simple reason that the past left something behind. That something is the material of the earth itself. With a few notable exceptions,[4] every atom of the earth's substance that is here in this instant will still be here a minute from now. It may change form or shape, it may move, it may even undergo a chemical reaction, but matter persists, and that means that the present always contains clues to the past.

A Fire in the Sky

An example of how the scientific method can be applied to an object well beyond our reach rises every morning in the eastern sky. Aside from the earth itself, no object in our tiny corner of the universe commands human attention and awe more than the sun. For many peoples, the sun was a God itself to be feared and worshiped. To the Egyptians he was Amon-Ra. To the Greeks he was drawn across the sky by Apollo in a chariot of gold. The list could go on, of course, punctuated by tales of mass panic as his rays were apparently lost at the height of a total solar eclipse. It is easy to see why our nearest star gained such status. The perceived movements of the sun define our days and nights, its warmth creates our seasons, and its energy grows the food that nourishes us and all of nature. To attribute all this to a power supernatural is understandable, maybe even logical.

If the sun's place in human imagination was once divine, its demotion to the status of mere matter surely began when Anaxagoras[5] argued in 434

B.C. that the sun was "just" a ball of fire floating in the air above the earth's surface. To be sure, Anaxagoras's analysis, which placed the sun a mere 4,000 miles above the surface of the earth, left a little to be desired. It seems that Anaxagoras had learned that on the day of the summer solstice the noonday sun in the city of Syene (the present-day location of the Aswan High Dam) cast no shadow—it was directly overhead. At exactly the same time, the sun in Alexandria, 500 miles to the north, cast a shadow of 7 degrees. Anaxagoras, no doubt a quick study in trigonometry, had taken these angular differences as a quick and easy way to calculate the height of the sun. It would have worked, too, if the earth had been flat. That little detail caused him to mistake the radius of the earth, which itself is close to 4,000 miles, for the distance to the sun.[6]

Despite the inadequacies of his analysis, it turns out that Anaxagoras's contemporaries were not at all amused by his calculations. He was condemned by the authorities and banned for life from the City of Athens, such was the rage of its citizens against the notion that the sun could be explained as mere matter.

Today we know a great deal more about the sun. Any high school student should be able to tell you that our sun is mostly hydrogen, and that it is driven by nuclear fusion reactions, which release vast quanities of energy that light our solar system and keep the spark of life burning on our little planet. It is only fair to ask how we actually know *anything* about the sun. No one has ever been there, no spacecraft has brought back a sample of solar plasma, no probe has ever landed on its fiery surface. So, from where does our knowledge of this nearby star come?

The answer is that, although we have never visited the sun, the sun's radiation does visit us. A careful analysis of that radiation provides us with all the clues we need to set Apollo firmly and conclusively into the realm of mythology. We know, for example, that the surface of the sun is much too hot to fit in anybody's chariot. How can we be sure? Laboratory experiments show that the spectum of light given off by superheated gases varies as a function of its temperature. The German physicist Wilhelm Wien, in fact, showed that the wavelength of maximum intensity given off by a hot body in the laboratory was inversely proportional to its temperature, and the single most intense wavelength of solar radiation is 480 nm (nanometers), which falls in the green region of the spectrum.[7] Wien's formula gives a surface temperature in the vicinity of 5,580 degrees Centigrade. The same principles, incidentally, can be used to determine the temperatures of other stars. All we need to do is to gather enough starlight to pass through a prism, and then to measure the relative intensities of the various wavelengths. So much for temperature.

Isaac Newton himself reported the ability of a prism to refract sunlight into its constituent colors. His simple observation held the key to another fundamental discovery about the nature of the sun. Newton's rainbow, produced when a broad spot of sunlight passes through a prism, is beautiful but imprecise. The reason is that individual wavelengths blend together, smeared across the spectrum by the diameter of the spot itself. Make the spot smaller, and the colors become purer. The ultimate, a single narrow slit, produces a broad, accurate spectrum in which each wavelength occupies a distinct position.

A century after Newton, German physicist Joseph Fraunhofer applied this technique to sunlight, and was stunned by what he saw. The solar spectrum was interrupted by hundreds of sharp, dark lines, wavelengths at which little or no light was present. By contrast, a laboratory object heated to 5,000 or 6,000 degrees Centigrade gave off a continuous spectrum, unmarred by what became known as "Fraunhofer lines." Was something wrong with the sun?

The sun, as it turns out, was just fine. A few decades later, Gustav Kirchoff passed white light from a high intensity source through a gas flame into which he had sprinkled a sodium compound, probably ordinary table salt. When he looked at the spectrum of the light coming through the flame, he saw a dark line that looked remarkably like a Fraunhofer line. He understood the laboratory result perfectly. The ability of an atom to emit or to absorb light depends upon the arrangement of electrons in its orbitals. Because each element has a different electron arrangement, each element absorbs light at a distinctive set of wavelengths. The dark line in his lab spectrum was produced because vaporized sodium atoms absorbed much of the light at a wavelength of 589 nanometers. Why was there an identical line in the solar spectrum? Kirchoff could think of only one reason. The white light produced by the sun must be passing through vaporized sodium. Where? In the atmosphere of the sun itself. The sun contains sodium.

Kirchoff's insight made it instantly possible to probe the solar spectrum for connections to each of the hundreds of Fraunhofer lines. Within a few decades much of the elemental composition of the sun was determined, including the important fact that it is more than ninety percent hydrogen. One slight mystery remained. There is a very faint Fraunhofer line close to one of the lines produced by hydrogen, so close that it was often mistaken for one of the hydrogen lines, and ignored. But in 1868, during a total solar eclipse, the English scientist Norman Lockyer recorded a spectrum from the solar corona. Such light is not filtered through the solar atmosphere, and the spectrum contained a distinctive emission peak at a wavelength of 587.6 nanometers. This was close to the peak for sodium, but not close enough. In fact, as Lockyer gradually came to realize, it did not match *any* terrestrial element.

Faced with such an unexpected result, Lockyer drew a bold but sensible conclusion. The sun contained a chemical element that was not found on earth. Drawing from the Greek word "*helios,*" he named it "helium." More than twenty-five years would pass before chemist William Ramsey would report the presence of helium on this planet. It may seem remarkable that the gas of party balloons and airships was once unknown, but the true wonder is that helium was first discovered at a distance of 93 million miles.

The abundance of solar helium, which makes up about eight percent of the sun, was a crucial clue to the secrets of the stars. Well before another century had passed, atomic physicists had shown that the solar fire was not chemical, but nuclear in nature. Helium was produced from hydrogen in a fusion reaction that took place when small nuclei were slammed together at high temperature and pressure. If this was true on the sun, it should also be true on the earth. It was. The hydrogen bomb, which employs an atomic explosion to generate those temperatures and pressures, uses exactly this principle. Ironically, an understanding of the sun's life-giving rays led directly to the development of a terrible weapon of war.

The Material of Understanding

Despite these achievements, no one would suggest that our understanding of the sun is now so complete that it is time for the practitioners of solar physics to close up shop. It would be foolish to suppose that there is no reason to explore the sun by more direct means. The next century may present opportunities to do exactly that. When that happens, it would not be surprising to find that our picture of the sun may have to undergo a few revisions. Properly understood, all scientific knowledge is tentative, and no answer is ever final. Still, there is every reason to believe that we do know enough about the sun to regard its location, its temperature, its chemical composition, and the source of its energy as scientific facts.

Can we really consider the composition of the sun, which we have never visited, as a scientific fact? A critic might point out that there is an assumption woven into the heart of the scientific method. That assumption goes by many names, but I propose we call it "scientific materialism."

Scientific materialism assumes that the objects and events of the natural world can be explained in terms of their material properties. When Kirchoff found that sodium in the laboratory produced a dark line identical to one of the Fraunhofer lines in the solar spectrum, he instinctively made use of that key assumption. If only one element can produce that line on earth, he reasoned, then only that very same element could produce it on the sun. By assuming that the laws of physics and chemistry are constant, Kirchoff and others extended the experimental reach of science all the way to the sun, 93 million miles away.

It's true that scientific materialism makes a considerable leap of faith. At its core is the belief that natural phenomena can be explained by material causes. That belief, of course, could be challenged. If I wanted to oppose the assumption of materialism, I might walk into a meeting of solar physicists, for example, and claim that the sun does not contain helium. Someone in the group would be likely to ask a simple question: "How, then, do you explain the 587.6 nanometer emission peak in the solar atmosphere?" My response: I do not have to explain it! Light from the sun, I would claim, is a miracle. Supernatural forces are responsible for that light, and such forces are beyond scientific explanation. "You'd might as well admit," I would insist, "that my explanation is just as good as yours. The only difference is that you pretend to be objective when you are not. Your so-called scientific work has a hidden, underlying bias in favor of scientific materialism. I have no such bias. Indeed, I'm the one with a truly open mind, because I can admit the possibility of the miraculous when you cannot."

Which is it? The miraculous or the material? In the case of the sun, the choice is obvious. Not only can Fraunhofer lines be duplicated in the laboratory, but analysis of the solar spectrum led to the discovery of an element in the sun long before its existence was ever suspected on the earth. When helium turned up in our own neighborhood, the implicit assumptions of scientific materialism were fulfilled. And the principal elements that seem to be on the sun are *exactly* those that would be expected if a long-term nuclear fusion process was taking place, the reality of which can be demonstrated in spectacular fashion by thermonuclear weapons. Does the "miraculous" explanation match such stirring persuasiveness?

Of course not. The statement that the workings of the sun are miraculous would place them beyond explanation, beyond investigation. If what happens on the sun is a miracle, then there is no point in trying to understand it. That line of reasoning is one of the things that has sometimes earned religion a bad reputation among scientists. If taken at face value, the miraculous explanation would tell us that science is not worth the trouble, that it will never yield the answers we seek, and that nature will forever be beyond all human understanding. Sterile and nonproductive in its consequences, the claim of miracle would put a lid on curiosity, experimentation, and the human creative imagination.

By contrast, as in the case of the sun, the results of scientific materialism merge into a tight and consistent web of theory and phenomena. Fortunately, no serious person today would invoke the supernatural to account for the warmth of the sun. The reason, of course, is that the scientific, materialist explanation *works*. The sun may be unimaginably distant, no indisputable sample of solar matter has ever been examined in the laboratory, and no working scientific probe has ever transmitted data from its surface. Yet it's still possible to figure out what's going on in the star next door. Science at a distance is still scientific. And that's exactly the point.

Pop-Top Science

Does this apply to the distance of history too? If the scientific method allows us to investigate the *distant,* does it also permit us to study the *ancient?* The answer, of course, is yes, and in many respects a more enthusiastic yes. In one important way, our links to the past are stronger and more informative than our links to astronomical neighbors, because the clues are more direct. The material of the past is still with us. If we know where to look, we can literally dig up what the past has left us, and examine it. That rule applies even if the past has left us nothing but garbage.

To choose a trivial example, consider the beer can. Some of my friends might regard the canning of beer as one of the great achievements of Western civilization. Whatever beer's virtues—and I believe they are considerable—one generally does not expect to see the beer can used in scientific study.

Having grown up in the postwar baby boom generation, I still remember the days when metal cans of beer and soft drinks were occasionally shaped like bottles, complete with classic bottletops. These containers, before they went extinct in the late 1950s, coexisted alongside more modern-looking cans with perfectly flat tops. Both types of cans required an opener, creating a market for a cheap, bifunctional tool with a bottle opener on one end and a can opener on the other. Generations will fondly remember the "church keys" they used to pop bottle tops and puncture cans. Dedicated beer drinkers, to avoid agitating the beverage as they poured it, would expertly puncture two holes in the top of the can. One allowed the golden liquid to pour out, and the other allowed air to flow in as the can was drained. Beer and soda vendors at ball games, for example, often carried heavy-duty industrial church keys that could punch both holes at the same time. This was a great system, but it did require that you keep a church key around. What happened to that ice-cold beer if you couldn't find the can opener? Nothing—unless you were so desperate you were willing to puncture the can with a screwdriver, a penknife, or even a rock.

Beverage manufacturers, ever eager to expand their sales by making their products easier to use, realized that the church key system was an impediment to consumption. In 1962,[8] they figured out a way to eliminate that impediment by devising a can top that did not need an opener. The "pop-top" can, as it was called, contained a premanufactured opening covered by an aluminum pull tab. A family could now bring a six-pack of soda on a picnic without worrying about whether or not they had also packed an opener. The pull tabs were so small that people generally discarded them on the spot, leading to a sudden littering of public places with slivers of aluminum. In some places the problem was considered so serious that local governments engineered campaigns to convince thirsty citizens that the only socially responsible thing to do with your pop-top was to drop it right back into the can.

Pop-tops, as it turns out, were only an intermediate stage in can engineering. In 1975, a new easy-open can was produced in which a metal lever is activated to press down and bend a tab back into the can itself. It opens easily, but now nothing breaks off from the can, and no litter is generated apart from the can itself. Although things have quieted down a bit since then, refinement of the design continues. In 1993, several beer companies introduced the "wide-mouth" can, designed to allow air to flow back into the can in a way that allows beer to be poured more smoothly. Less foam and easier to chug.

The saga of the beer can might be of little interest to science, except for our propensity to throw them away. For decades, discarded beer and soft drink cans made up a large and consistent fraction of municipal garbage. That garbage was often buried in landfills or, in coastal regions, dumped into shallow waters offshore. To archeologists, garbage is a gold mine of information. Tight packing ensures good preservation, and the composition of well-preserved garbage can tell us what the people of a civilization ate, drank, wore, and used in their daily lives. You do, of course, have to be sure of what you are looking at before you can draw conclusions; and believe it or not, that's where the beer can comes in. As offshore garbage piles up, it is mixed in with other sediments and gradually buried deeper and deeper with time. Municipalities often changed their dumping locations without leaving records, so one cannot be completely certain whether a rich vein of garbage was dumped in 1968 or 1978. Until, that is, the beer cans are checked.

It turns out that the presence (or absence) of removable pop-tops, ubiquitous only in garbage discarded from 1962 through 1975, unambiguously identifies a narrow era in the modern archeology of trash. Investigators can study that particular moment in time, grateful for every beer-drinking couch potato who tossed his pop-tops in the can, time-stamping his refuse for future generations of science. To be sure, the inelegance of a corps of students sifting through dredged-up garbage may not fit our preconceptions of science, but it serves as a perfect model for how to investigate the past. Exhume materials created or deposited at a particular point in history. Identify that point in time by using a verifiable test—

in this case, the pop-top test. Then base your analysis of history on exactly what you find. The fact that we can analyze the materials of the past using the tools of today means that the past, within limits[9], is reconstructable. And the method of choice is scientific materialism.

Historians try to answer two classic questions: What happened in the past? and How did it happen? This is also true for paleontologists, who are our natural historians. Unlike beer cans, most living things do not leave traces of their existence. Only a few are buried quickly enough to escape the recycling processes of predators and decomposers, and these few then undergo chemical changes that alter their composition. However, living things are so numerous that those few that are left constitute a wealth of fossils, remnants of partially preserved forms that tell us much about the life of the past. And like a beer can, each fossil is a scientific fact that can be used to reconstruct the ebb and flow of life over time. The past is not only scientifically approachable, it is all around us, waiting to be dug up, investigated, and explored.

Looking Backwards

Against the backdrop of natural history, evolution is really a story of logic extended backwards into time. We begin to reconstruct the story by asking the first, and most important, question about the life of the past—was it the same as life today? The answer to that question is an unequivocal no. In an age of Jurassic Park and science fiction time travel, most of us will take that answer for granted; but throughout most of human history, our cultures and institutions were built on the assumption that the earth is an unchanging place.

Our ancestors can certainly be forgiven for making that assumption. There is a comforting certainty in the cycles of time, and as summer returns year after year, a steady ebb and flow of species marks the passage of each season. Seeing the regularity of birth, death, and migration, it would be easy to conclude that such has been the pattern of life for all eternity. But the seventy years that mark a human life span do not even approach the time required to detect life's true pattern on this planet. To find that, we have to look beneath our feet and read the record of the past.

I like to tell my students that if they wish to study the origin of evolution as a scientific idea, they should not begin with the life of Charles Darwin.

Rather, they should research another Englishman not generally considered to be a biologist—James Watt, the inventor of the steam engine. Watt's device helped to usher in the Industrial Revolution, but it also had a powerful secondary effect on the study of natural history. For the first time, hillsides were cut away by great machines mining coal and leveling for railways. The power of industry led to the first systematic, mechanized excavation of the landscape, and in so doing exposed great volumes of earth and rock to the eye of man. Suddenly, fossils, those once rare, lifelike impressions of strange organisms, were everywhere to be found. And not at all unlike that succession of beer cans buried in municipal waste, they had a tale to tell. Thank you, James Watt!

We could recount this tale in any one of a number of ways. We could begin with the ancients, and point out that fossils were described by the likes of Herodotus, Plutarch, and Xenophanes. We could recount explanations for these lifelike impressions of strange animals. Aristotle explained fossils as being the result of shifting sea levels that had trapped and preserved the shells and skeletons of small marine animals. Others suggested that fossils simply grew out of nonliving rocks, mere accidents of nature. Still others maintained that all fossils were the remains of organisms that had perished in the Flood of Noah. That view, incidentally, held sway despite the ingenious demonstrations of Leonardo da Vinci that fossils had not been formed simultaneously. Or we could visit southwestern England in 1794, watching a young man named William Smith supervising the construction of a ditch that was to become known as the Somerset Canal.

Unlike the gentleman naturalists of the day, Smith was a person of modest means from a family of farmers. His fortunes were lifted in the great tide of ingenuity known as the Industrial Revolution. Born in Churchill, England, in 1769, Smith learned the practical arts of mapping and surveying. His successes as an apprentice to well-known surveyor Edward Webb were notable, and he soon found his civil engineering skills in great demand as roads and canals began to crisscross Britain. At the age of twenty-five he was placed in charge of the massive Somerset Canal project. Surveying the way of this canal was his principal task, but he quickly discovered that the early stages of construction required careful attention to the location through which a canal would be cut. Porous rock, which might drain the canal's water, would obviously be a poor choice. Smith was forced to master practical geology on the job, and master it he did.

The great cuts of earth required to build the Somerset Canal exposed layer after layer of geological strata. Smith soon came to realize that rock formations often appeared in a regular order, a pattern that might allow him to predict from surface rocks whether a downward cut of fifteen or twenty feet would expose solid bedrock or the more porous formations he hoped to avoid. Earlier geologists had also noticed such regular patterns, but local variations in rock composition and color had made it impossible for them to generalize over distances of the sort that Smith's project was to cover. Fortunately, he found another clue:

> Each stratum contained organized fossils peculiar to itself, and might, in cases otherwise doubtful, be recognised and discriminated from others like it, but in a different part of the series, by examination of them.[10]

In other words, Smith learned that he could identify individual layers of rock by fossils unique to each layer. His notebooks show that he seized upon this observation early in the Somerset project. On January 5, 1796, he wrote:

> Fossils have long been studied as great curiosities, collected with great pains, treasured with great care and at a great expense, and shown and admired with as much pleasure as a child's hobby-horse is shown and admired by himself and his playfellows, because it is pretty; and this has been done by thousands who have never paid the least regard to that wonderful order and regularity with which nature has disposed of these singular productions, and assigned to each class its peculiar stratum.[11]

Geologists will recognize in these words what they now call the principle of "faunal succession," the idea that fossils are laid down in a pattern that serves as an index to living history. When the Somerset Canal was finished in 1799, Smith set out to produce a complete geologic map of England and Wales, a work that was published in 1815 only with the help of hundreds of subscribers. By that time, Smith's use of fossils to trace geologic history had found wide application in France and Germany, and the message was clear. In the most recent fossils, ones recovered from the uppermost layers of sedimentary rocks, naturalists could recognize organisms nearly identical to those of the present day. But as they went deeper, they found differences, some slight, some profound. They even discovered the remains of organisms so unusual that they clearly had long since vanished from the face of the earth. This fossil record, for all of its imperfections, told an unmistakable story—life had changed over time, changed dramatically.

The new technology of steam and steel may have ravaged the landscape in search of fuel and speedy transport, but it also opened the door to our past.

Thinking the Unthinkable

Smith's insights on the fossil record led to the unavoidable conclusion that some organisms had become extinct and new ones had appeared to take their place. Such reasoning may seem pedestrian to us, but two hundred years ago the very notion of extinction was unacceptable to many. Some thought it blasphemous to suggest that *any* of God's creatures could have failed the test of time by vanishing into extinction. No less an authority than Thomas Jefferson believed that the Creator would never have allowed one of his creatures to disappear forever. The fossilized remains of unfamiliar organisms, Jefferson suggested, are simply those of creatures we have yet to discover in our explorations of the planet. Jefferson's view was not compelling to those who had traveled and collected more widely than he had.

In France, the great naturalist Georges Cuvier had already found a way to settle the question, mixing exploration and analysis in a way that set the standard for the new science of paleontology. Unlike Smith, Cuvier was able to devote his full time and attention to natural philosophy. He carried out detailed studies of elephant anatomy that showed, for the first time, that Indian and African elephants were distinct and different species. Then, using newly discovered fossil bones from Russia and Europe, he showed that the woolly mammoth was itself a unique species, different from any living organism, as was the American mastodon.

Cuvier went on to document scores of extinctions:

> Numberless living beings have been the victims of these catastrophes. . . . Their races have even become extinct, and have left no memorial of them except some fragment which the naturalist can scarcely recognize.[12]

So widespread were the losses of species that Cuvier came to describe these great extinctions as "revolutions" in the history of life. He was such an able anatomist, he cultivated the self-serving suggestion that he might reconstruct an entire skeleton from just one bone. Yet Cuvier was at a loss to explain how these great extinctions might fit into a connected pattern of natural history, a narrative that might provide causes for the faunal succession that was so apparent in the fossil record.

Nowhere was this more obvious than in a brief controversy with his colleague, Etienne Geoffroy Saint-Hilaire, on the unlikely issue of crocodiles. In 1823, fossil specimens of what came to be known as the "crocodile of Caen" were shipped to Cuvier for examination. A year later, he wrote up a report describing the bones as belonging to an extinct species of gavial,[13] a relative of the crocodile found today on the Indian subcontinent. Geoffroy, an anatomist of great skill, rexamined the bones, and before long realized that Cuvier's verdict had been too hasty. Geoffroy's more careful studies showed that the crocodile of Caen was so different from all known crocodilians that it could be placed in a genus of its own, and that is exactly what Geoffroy decided to do. He called it *Teleosaurus,* a name it still carries today. A related fossil, likewise distinct from modern forms, was given the name *Steneosaurus*. Although Cuvier would later carry the day in a much more important debate on the nature of vertebrate and invertebrate body plans, on this matter Geoffroy was one who got it right.

Geoffroy's little triumph over Cuvier might be forgotten today except for the fact that he also decided to inquire "concerning the degree of probability that the teleosauruses and the steneosauruses, animals of antediluvian ages, are the source of the crocodiles dispersed today in the warm climates of the two continents."[14] In other words, Geoffroy wondered whether or not these fossils might be the ancestors of present-day species. He wrote:

> It is not repugnant to reason, that is to physiological principles, that the crocodiles of the present epoch could be descended through an uninterrupted succession from the antediluvian species.[15]

The reasons for his speculation grew out of the very same anatomical studies required to win his conflict with Cuvier. Having established the unique characteristics of a variety of extinct forms, he was in a perfect position to appreciate the ways in which the subtleties of their anatomies anticipated the adaptations of present-day forms. In other words, the more he studied the uniqueness of the past, the more he saw within its forms the roots of the present.

Having proposed the unthinkable—that one species might give rise to another—Geoffroy realized that he had better come up with a mechanism to accomplish the transformation. On this score he was less successful. Geoffroy struggled to find a way in which the environment could shape and direct animal development. He worked on the assumption that forces acting during embryonic development could produce lasting changes in the form of an animal, and suggested that this might be the way in which the crocodile of Caen had produced its distinctly modern descendants. In so doing, he came remarkably close to describing what we would recognize today as evolution:

> The external world is all-powerful in alteration of the form of organized bodies . . . these [modifications] are inherited, and they influence the rest of the organization of the animal, because if these modifications lead to injurious effects the animals which exhibit them perish and are replaced by others of a somewhat different form, a form changed so as to be adapted to the new environment.[16]

Close—very close, in fact. But no cigar. The importance of natural selection would be left for Darwin to discover. Despite Geoffroy's lack of success in finding a mechanism to produce these "alterations," his willingness to make the obvious connection between extinct and living organisms changed forever the way in which fossils would be understood. It was now possible to find the roots of the present in the unmistakable records of the past.

Steps in the Parade

It is often, but quite mistakenly, believed that Charles Darwin was the first natural scientist to notice the pattern of evolutionary charge that is etched into the solid record of historical geology. As we have just seen, Smith, Cuvier, and Geoffroy, along with many others, had come to appreciate the depth and power of that record of change long before Darwin began to dabble in natural history. In other words, the story of change predates Darwin's theory of evolution. Evolution is partly the story of how the present is linked to the past, the story of *what happened*. In this sense, evolution is history.

As with all histories, our knowledge of the biological past is incomplete. We can reconstruct, analyze, and discuss much of the historical past with remarkable precision. Our understanding of U.S. history, for example, gets better with each new document, artifact, and photograph that is added to it; but the utter impossibility of knowing *everything* that happened to *everyone* means that any history will never be complete. The same is true for the history, the evolution, of life.

It is beyond the scope of this book to present anything close to a complete account of the history of life. To do so, we would have to trace the changing pattern of time throughout the classic geological ages of the earth's past—Precambrian to Cambrian, Cambrian to Ordovician, Silurian, Devonian, Mississippian, and so forth, all the way to the Quaternary period, which includes the present.

We could replay that history here, documenting the transformations, the extinctions, and the appearances that characterize the history of life. For good measure, we might just as well attempt to retell all of human history, tracing the developments of all languages, civilizations, cultures, inventions, and conquests.

Fortunately, as any sane historian would argue, you don't have to retell *all* histories in order to get a sense of what history is really like. Whenever new cultures and civilizations appear, as happens repeatedly in human history, we can find their roots in the immediate past. To some, the great American civilization that sprung into dominance in the twentieth century may seem to have seized the world stage out of nowhere, but students of American history know better. They understand the unique conditions that first infused this continent with a critical mixture of English law and tradition. They would trace the transformation of that structured society by the demands of a frontier that leveled the assumptions and obligations of privilege. Ultimately, they would find the roots of today's American civilization in a score of European tribes, in the captured sons and daughters of Africa, in waves of hopeful immigrants from Asia and the Americas, and even in the surviving civilizations of the country's native inhabitants. In short, they would find today's America, however new and novel, linked historically to the pasts of dozens of other civilizations.

What the history of this one remarkable nation tells us about history in general is that we can understand the present only in terms of a detailed study of the past. This idea—more important than we often appreciate—is the basis for the systematic study of numan history.

The same principle is at the heart of every study of the biological past. Just as human history reveals a succession of languages, civilizations, and empires, natural history reveals a succession of living organisms that are linked in a stunning pattern of relatedness.

Even an elementary chart (Figure 2.1) of the earliest appearances of groups of organisms shows this pattern. Fossils containing the earliest evidences of life appear in the Proterozoic ("before life") era, and many are more than 3 billion years old. Fossils formed in this era contain clear microscopic impressions of prokaryotic cells, cells that (like modern bacteria) do not have nuclei. For the next 3 billion years, most of the history of life on earth, that's all there was. Then, about 1.5 billion years ago, the first of a series of new microscopic fossils appear that have traces of what may be nuclei within their cells. The eukaryotic cell, with genetic information contained within the nucleus, had finally appeared.

Having taken almost 2 billion years to gain a nucleus, life took several hundred million more to figure out how to become multicellular. The first widespread experiments in this regard were a stunning series of multicellular animals fossilized almost 700 million years ago into rock that now is part of the Ediacaran Hills of Australia. These remarkable organisms may or may not be directly related to animals alive today, and paleontologists disagree as to whether the Ediacaran fauna were actually dead ends or included the founding species of more successful and enduring lines.

The next well-documented period, the Cambrian, clearly does contain a number of organisms ancestral to the great phyla of modern animals. From this point on, new groups of organisms appear in a pattern of historical succession that is very well documented. Shellfish and corals, for example, date to the Cambrian, more than 550 million years ago. The first true fish appear in the Ordovician, about 480 million years ago. Amphibians appear 380 million years ago, the first reptiles 40 million years later, and the first true dinosaurs nearly 80 million years later in the Triassic. The earliest true mammals show up around 210 million years ago; the first birds appear late in the Jurassic, 155 millon years ago; and the last of the

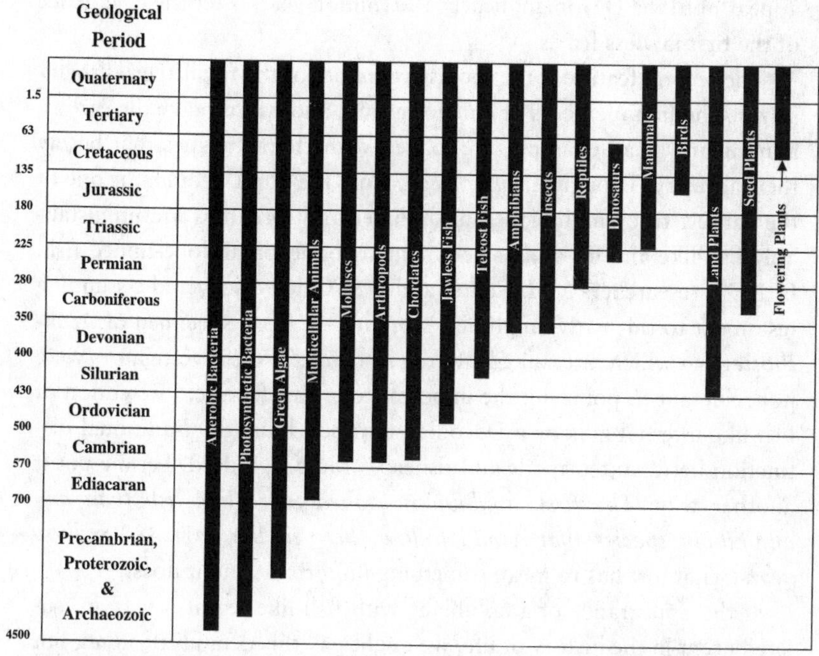

Figure 2.1. The appearance and persistence of major groups of organisms over geologic time. The beginning of each period is shown in millions of years before present.[17]

dinosaurs vanish by extinction at the end of the Cretaceous, 65 million years ago.

Two important features of this familiar recounting of natural history are crucial to understanding the power and scope of the fossil record. The first is that the emergence of a major group is not the end but the beginning of a new round of evolutionary change within that group. Those early fish from the Ordovician, to take a well-preserved example, were organisms covered with thick scales but lacking jaws or bones. Primitive jaws appeared gradually over the next two geological periods, produced by a gradual modification of the first two gill arches to produce a structure that could open and close the mouth at will—possibly to improve respiratory efficiency. Still to come were improved jaws and skeletons. Fish with true bones, which dominate the seas today, did not appear until the Devonian, nearly 100 million years after the emergence of the first jawless fishes.

The second feature is the successive nature of the fossil species themselves. The first members of a major group, and we can take the earliest amphibians as an example, are loaded with characteristics that betray their ancestry. Though *Acanthostega,* from the late Devonian, is one of the earliest true amphibians, its shoulder and forelimbs are unmistakably fish-like and its skull is very similar to Devonian lobe-finned fish. In 1991, researchers M. I. Coates and J. A. Clack discovered a stunning testament to this early amphibian's origins—a fossil specimen of *Acanthostega* in which internal gills were preserved. These investigators were not reluctant to point out the importance of this feature: "Retention of fish-like internal gills by a Devonian tetrapod blurs the traditional distinction between tetrapods and fishes."[18] Blurs it, indeed. Let me put it another way: *The first amphibians looked more like fish than any amphibian species that would follow them in the next 380 million years*. That just has to mean something important, and it does.[19]

If the appearance of amphibians with fish-like features was an isolated event in the history of life, we might pay this detail little mind, but that is not the case. The first reptiles to appear in the fossil record are more amphibian-like than any reptiles to follow. The first mammals have a set of reptilian characteristics so pronounced that they are commonly known as the reptile-like mammals. The first birds are so similar to another group of reptiles that some paleontologists have formally proposed that birds be classified as a subgroup of the dinosaurs.

Time after time, species after species, the greater our knowledge of the earth's natural history, the greater the number of examples in which the appearance of a new species can be linked directly to a similar species preceeding it in time. These histories reveal a pattern of change, a pattern that Darwin aptly called not "evolution," but "descent with modification." Once this pattern becomes clear, and it can be found in any part of the fossil record, the theme of life is equally clear.

Patterns in Place and Time

Skepticism is a cardinal virtue in science. If the sequential character of the fossil record is genuine, and not the prejudgment of imagination constrained to view natural history through a Darwinian lens, it should be able to meet the test of place as well as the test of time. In other words, if the multitude of fossil sequences really are linked by ancestor-descendant relationships, we should be able to detect geographic patterns as well as temporal ones in the fossil record. And we do.

Much of the scholarship on Darwin suggests that this fact dawned on him slowly. Adrian Desmond and James Moore's biography of Darwin[20] describes a young naturalist who didn't quite realize the importance of the trove of fossil bones he had brought back from South America. He was not keen enough to identify many of these specimens, some of which were large enough that he might well have thought them to be the bones of rhinos or mastodons. With the assistance of geologist Charles Lyell, Darwin passed many of these specimens to the talented and ambitious anatomist, Richard Owen. Owen's considerable skills enabled him to establish the true identity of the specimens.

After a few weeks of study, Owen came back with a shocking conclusion for the young Darwin. The first fossil he investigated was the skull of a hippo-sized organism that Owen named *Toxodon* (Figure 2.2). In Owen's view, the detailed anatomy of its skull showed that *Toxodon,* despite its great size, was a rodent.

Darwin and Owen may have assumed it was related to the capybara, the largest rodent alive today.[21] As Owen's work continued, he identified a slew of organisms from Darwin's trove. They included giant llamas, predatory cats, and rabbit-like marsupials. One of his prize specimens was a huge, cow-sized armored mammal given the name *Glyptodon* (Figure 2.3) Despite its mass, its skeletal similarities to a modern American mammal, the armadillo, were unmistakable.

The importance of these findings escaped neither Darwin nor his mentor, Lyell. South America had proven to contain a fauna that was strikingly unique, boasting animals found nowhere else on earth. To drive the point home, South America also contained a fossil record packed with those animals' closest anatomical relatives. Why, one might ask, should such a unique set of animals be found in exactly the same place as their closest fossil relatives? There could be just one answer—a process of descent with modification linked the present to the past. If the armadillo, to take just one example, was a species that had been created *ex nihilo,* why was it found in the exact same spot on the globe as the fossil species most similar

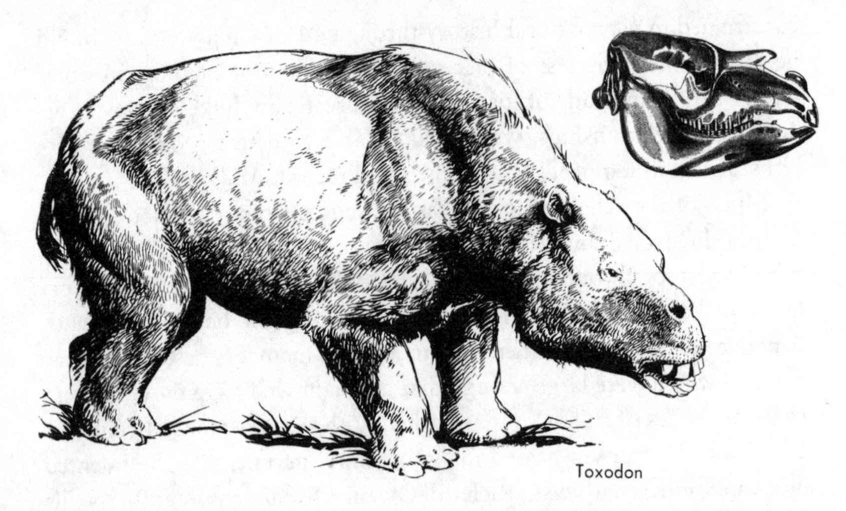

Figure 2.2. Skull and an artist's reconstruction of *Toxodon,* one of the fossil prizes from Darwin's trips to the South American mainland.

Figure 2.3. *Glyptodon,* a huge armored mammal related to the modern armadillo. The striking similarities observed by Darwin between South America's ancient and modern fauna suggested an ancestor-descendent relationship between past and present.

to it? The answer, of course, is that the armadillo was found only in the New World because it had *evolved* there from its ancient ancestors.

Exactly the same considerations can be applied to the fossil animals of North America, Eurasia, Africa, and especially Australia. In each and every case, the pattern of temporal succession of the fossil record is linked to a matching pattern of geographical succession. The geographically unique species of today are linked in time to the unique species of the past. The bottom line is that descent with modification, which most of us prefer to call evolution, really happened.

Change in Detail

If evolution is *what* happened in the past, the next logical question to consider is *how* it happened. Many of his critics have written that Darwin was vague on the mechanism of evolution, and that despite the title of his great work, he had little to say on the actual process by which species originate. I disagree.

Darwin not only addressed the formation of new species in *On the Origin of Species,* he made speciation a focal point of his arguments.

So that the reader could be sure what he had in mind, Darwin went to the great trouble of drafting a single figure to include in his book, illustrating the divergence of species over time (Figure 2.4).[22]

As he wrote in the summary to his fourth chapter, the range of variety observed within a species is already enough to get the process going:

> Thus the small differences distinguishing varieties of the same species, steadily tend to increase, till they equal the greater differences between species of the same genus, or even of distinct genera.[23]

It's only fair to put Darwin's account of evolutionary change to the simple test of observation. Do we see evidence in the fossil record for the kind of change he sketched in his diagram? And can we see evidence of species today that sit at the branching-points he considered critical to the origins of new species? The answer to each question is an unequivocal yes.

The fossil record first. As Darwin himself observed, the fossil record is terribly incomplete. Preservation is haphazard, and is strongly skewed in favor of marine organisms. The art of finding anything is the art of knowing where to look. In the century and a half following Darwin, paleontology has found plenty of good places to look. Here I present two examples of speciation, one chosen because the record is almost complete, the other because it's a particularly interesting species.

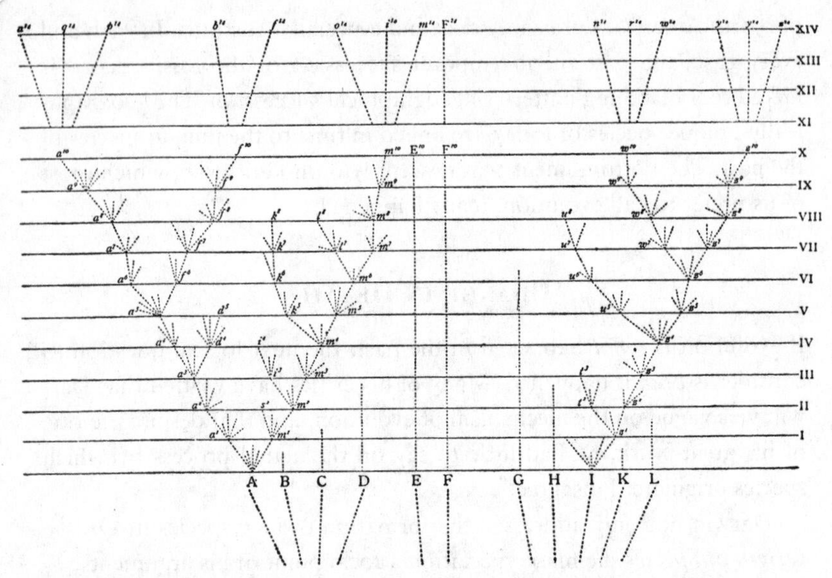

Figure 2.4. Darwin's diagram showing the "Divergence of Taxa" from *On the Origin of Species*. According to Darwin, "A" to "L" represent the "species of a genus large in its own country." Taking species A as an example, the "fan of diverging dotted lines of unequal lengths proceeding from (A), may represent its varying offspring." When "a dotted line reaches one of the horizontal lines," enough variation has accumulated that distinct varieties have formed. After "ten thousand generations, species (A) is supposed to have produced three forms, a10, f10, and m10," each of which may now be distinct enough to be considered separate species or subspecies. This drawing, the only illustration included in *The Origin,* shows speciation as well as extinction, evolutionary stasis, and the uneven pace of evolutionary change.

Rhizosolenia is a genus of diatoms, single-celled photosynthetic organisms that produce intricate and distinctive silicate cell walls. When their owners die, these glass-like walls fall to the ocean floor and produce a sediment so dense that it can be mined as diatomaceous earth (and used as a filtering matrix for swimming pools). Two distinct species, *Rhizosolenia praebergonii* and *Rhizosolenia bergonii,* are known from sediments dating to 1.7 million years ago. If we trace these species backwards in time, we gather data (Figure 2.5) that duplicate, with uncanny precision, Darwin's drawing of speciation.

Beginning at 3.3 million years before the present, we can see the increasing range of diversity of the ancestral species, leading to a broadening at 2.9 million years that splits into two distinct lineages in less than 200,000 years. The continuous deposition of diatom shells has provided a complete record covering nearly 2 million years. Thomas Cronin and Cynthia Schneider, who reported this study, were lucky enough to find a speciation event right in the middle of their data.[24]

Most paleontologists are not lucky enough to find an abundance of well-preserved fossils exactly straddling both sides of the time point when a new species was formed, but the same pattern seems to exist even where the fossil evidence is not as abundant.

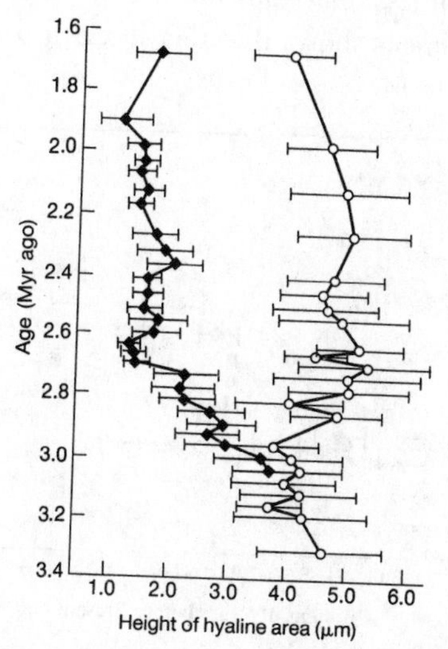

Figure 2.5. In those rare cases where a nearly complete fossil record is available, as with the diatom *Rhizosolenia,* Darwin's ideas on the divergence of species match empirical data remarkably well. The split of a single species into two (*R. praebergonii* and *R. bergonii*) is shown as a function of the hyaline (glass-like) area of the cell wall. The speciation event is first detectable at 3.05 million years before present, and is complete within 200,000 years.

Figure 2.5 is adapted, with permission, from Thomas M. Cronin and Cynthia E. Schneider's "Climatic Influences on Species: Evidence from the Fossil Record," *Trends in Evolutionary Biology and Ecology* 5 (1990): 275–279. The original data shown in the figure were from D. Lazarus, *Paleobiology* 12 (1986): 175–189.

Take as a case in point a certain type of mammal believed to have originated in Africa during the past few million years. The fact that we are looking in the recent past is in our favor, but this particular group of organisms lived on land, sometimes in trees, and therefore rarely died in a way that would leave fossils. Since we are *really* interested in these organisms, we press ahead with what we have.

After decades of work, we can take one of this organism's most distinctive physical characteristics—the size of the most anterior region of its nervous system—and graph it over time in the same way in which we plotted the height of a portion of the *Rhizosolenia* cell wall. Figure 2.6 shows the result.

An ancestral species dating back to 3 million years before present has a brain size of somewhat less than 400 cubic centimeters. The discovery of many individual specimens shows that this species persisted until some-

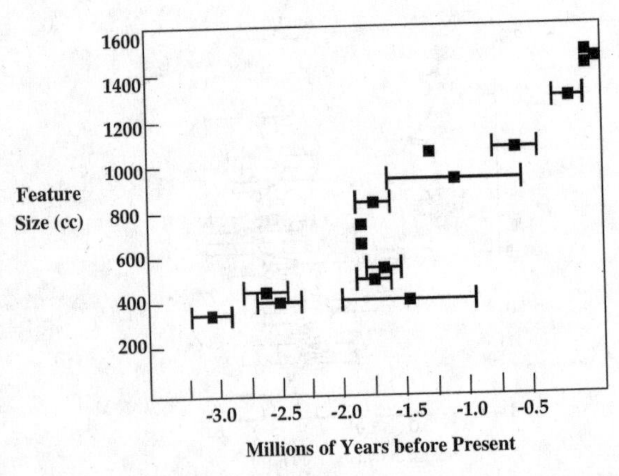

Figure 2.6. Another representative case of species divergence as documented in the fossil record. Although the fossil record here is not nearly as complete as that of *Rhizosolenia*, the same pattern prevails. This figure documents the evolution of our own species, *Homo sapiens*, and the feature sizes plotted as against time before present are the cranial capacities of individual fossils in cubic centimeters (cc). Fossils with cranial capacities of less than 600 cc are generally placed in the genus *Australopithecus*, and those greater than 600 cc are placed in our own genus, *Homo*.[25]

Redrawn from a figure included in Dean Faulk's article "Hominid Brain Evolution: Looks Can Be Deceiving," in *Science* 280 (1998): 1714.

where between 1 and 1.5 million years ago. About 2 million years ago, however, the diversity of the population spread so widely that the species split in two, giving rise to a new, but closely related, species. In this new branch, brain size increased smoothly from 650 cc to 1,500 cc over 2 million years. Briefly, as the two lines split, the two species coexisted much like *R. praebergonii* and *R. bergonii*. Before long, the ancestral line vanished, and only the larger-brained lineage continued to the present day.

The data in this figure represent the evolution of our own species from *Australopithecus*. For all the fuss and concern that surround the idea of human evolution, the detailed fossil evidence of our ancestry is remarkably powerful. The origins of our species fit Darwin's sketch just as surely as do the origins of diatoms. The record of change fits the pattern predicted by evolution, and evolution produced us, too.

Can we find, in the present-day world around us, species whose variability and distribution match the branch-points of Darwinian speciation? Once again, the answer is yes.

The herring gull (*Larus argenatus*) and the lesser black-backed gull (*Larus fuscus*) are both found in northern Europe, and they meet the usual tests of distinct species. The birds are distinctive in color, have different markings on their legs, nest in different sites, and do not interbreed. Even an amateur bird-watcher has no trouble telling the two apart. If we move to North America, where only the herring gull is found, and then begin to travel westward along the polar latitudes, something strange begins to happen. As we go farther west, from Alaska to Siberia, to northern Europe, the coloration of the herring gull changes. It gets darker and darker, until at a point in central Siberia it becomes so dark that it can be classified as the black-backed gull.

These birds are just one well-known example of what is called a "ring species," a group of organisms whose extremes of variations easily meet the test of distinct species, but whose extremes are linked by intermediates. All that would be needed to complete the split into two species would be the extinction or isolation of those intermediates. And that could happen as a result of climate change, predation, disease, or natural catastrophe.

Closer to home is a spectacular ring of salamander species (or incipient species) that surrounds the Central Valley of California. David Wake of the University of California has described as many as seven distinct varieties that make up this ring, each different enough from its neighbors to merit distinction as a separate species. Comes the next earthquake, brushfire, or freeway project, that is exactly what is likely to happen. Evolution, like life itself (and like highway construction), continues.

Evolution as a Creative Force

However detailed the fossil record becomes—and scarcely a month has gone by in the last few years without an important discovery to make that record more complete—there will always be those who argue that the historical record of life is not really the point. The issue, they might say, is the *mechanism* that drives that process of change. Evolution's opponents would argue that the Darwinian mechanism of variation and natural selection is not sufficient to produce the novel structures and new morphologies that appear, even over millions of years, in the natural history of life.

This is a point that I will return to in the pages ahead when we consider whether or not evolutionary mechanisms are capable of producing the biochemical complexity of the cell. For now, let's answer the argument in the simplest way possible. What *can* evolution do?

Even the opponents of evolution agree that natural selection is a genuine force in shaping the characteristics of organisms, generation after generation. Favorable variations increase the likelihood of success in the struggle for existence, and therefore natural selection automatically chooses those characteristics best suited for survival and weeds out those that are least helpful. So far so good, evolution's critics would say, for the ability of selection to choose from existing variation, but it is the source of new variation that falls short. For Darwin's mechanism to work, to produce evolutionary novelty and long-term biological change, new variation must arise constantly in the form of genes that change the nature and improve the fitness of organisms, from grapefruits to wombats. This, they say, is impossible.

As a researcher in the field of cell biology, I have to admit that my first reaction to such claims has always been puzzlement. Living things, after all, are constructed by the execution of a series of genetic messages encoded in DNA. Genes, the functional units of that genetic program, generally encode proteins, which are the workhorses of the cell. As our exploration of the genomes of humans and other organisms expands, it becomes clear that those proteins can do just about everything required to produce an organism—the types of proteins produced in an embryo determine the location of its head and tail, its right and left. They determine which cells will produce an eye, and which will grow into a limb. If proteins can do all of this, then mutations, spontaneous changes in the genes that encode these proteins, can change each and every one of these characteristics. Mutations are a continuing and inexhaustable source of variation, and they provide the raw material that is shaped by natural selection. Since mutations can duplicate, delete, invert, and rewrite any part of the genetic system in any organism, they can produce any change that evolution has documented.

The opponents of evolution never deny that mutations produce variation, but they do argue that mutations, being unpredictable in their effects and random in their occurrence, cannot produce beneficial improvements for natural selection to work upon. In short, mutations just mess things up.

As an experimental biologist, I am inclined to look for an empirical test. Can we place an organism in a situation where its ability to generate random beneficial mutations will be tested? The answer is yes, and we do exactly this every day. The results of such tests have been known for years; and yes, evolution can generate lots of beneficial mutations—so many, in fact, that we can say that the evolutionary mechanism is not only real, but a downright nuisance.

In the 1940s, physicians were given a dramatic new weapon in their fight against bacterial infection—the antibiotic penicillin. Penicillin kills bacteria in a particularly sneaky way. Most bacteria are surrounded by a tough cell wall that protects them from swelling as water tries to diffuse across their cell membranes.[26] As they grow, bacteria must expand and remodel that cell wall, which they do gradually by making microscopic openings in the wall, adding strand after strand of microscopic fibers, then cross-linking those fibers to cement the new wall together.

Penicillin inhibits the enzyme that does the cross-linking. As a result, when *Staphylococcus* or *Streptococcus* bacteria grow in the presence of penicillin, they produce cell walls that are dramatically weakened by the absence of new cross-links. When they reach a certain point, the pressure caused by the inward movement of water becomes more than the shoddy wall can stand. The wall breaks, the bacterium expands like a balloon, and bursts. Score one for modern medicine.

But bacteria are able to fight back. The result of flooding hospital wards and animal feedlots with penicillin and similar compounds has been the emergence of bacteria strains that are partially or totally resistant to antibiotics. Score one for the bad guys. In fact, the emergence of drug resistance has been a major factor in the spread of a variety of diseases, especially tuberculosis. The culprit is evolution. By creating conditions under which only resistant bacteria will survive, we have produced a new kind of natural selection in which mutations that confer antibiotic resistance are beneficial to the bacterium. The result is unavoidable, given the millions of genetic duplications that occur in a bacterial population in just a few days. Sooner or later, the "right" mutation shows up, and it causes the individual bacteria that possess it to prosper at our expense.

Similar problems crop up whenever we try to control another species with chemical means, as farmers dependent upon pesticides can readily confirm. The opponents of evolution sometimes complain that resistance genes for these compounds already exist in the population. They contend that these examples show how resistance can be selected for, not how it is produced in the first place. If that were the case, permanent victory over such organisms could be achieved by engineering a synthetic compound that had never before appeared in nature. Since there could be no preexisting resistance to such a brand-new, manufactured compound, evolution, in their view, could never produce resistance to it. Many medical scientists might wish that this were the case.

Unfortunately, it is not. In 1996, a new class of drugs, produced by just such a process, were specifically engineered to block HIV-protease, one of the key enzymes used by the virus that causes AIDS. Compounds unlike any found in nature were engineered to fit precisely into a crevice in this protein enzyme, to block its activity. HIV-infected people treated with these new protease inhibitors showed dramatic improvements. The drugs blocked viral replication and delayed the onset of symptoms so effectively that AIDS-related hospitalizations dropped dramatically. A new age of therapy seemed to be dawning.

However, evolution enabled HIV to strike back. The virus has a very high mutation rate, owing to the sloppy work done by the reverse transcriptase enzyme it uses to copy its genetic material. During the course of long months and years of treatment, mutations in the HIV-protease gene appeared in the very bodies of people using the drugs. Gradually the effectiveness of the drugs waned, owing to the emergence of new, drug-resistant varieties of the virus. Why were these viruses drug-resistant? Because they had undergone mutations that remodeled their proteases, enabling them to do their work without allowing the protease inhibitors to block them. In short, that random, undirected process of mutation had produced the "right" kind of variation for natural selection to act upon, even within the body of one individual. Evolution may have made us, but that does not mean that it is always our friend. It made the bad guys, too.

Evolution as a Tool

I believe that one of the things that bothers people most about evolution is the simplicity of its three-part mechanism. Mutation, variation, and natural selection. Is that *really* all there is to it? "Well, yes," is the proper answer, followed by a very quick disclaimer. "There's a lot more to that mechanism than meets the eye." A lot more.

A news report in *Science* magazine a few years ago hinted at the surprising power of this modest mechanism. A reporter interviewing Gregory Petsko of Brandeis University recounted the trouble Petsko's lab had encountered in trying to get a bacterium's genetic machinery to increase the output of a particular protein. "We went to a lot of trouble," he said, without getting good yields. But then a chance event in just one cell (out of millions) produced exactly the variation they needed. The cell "took our gene and moved it thousands of bases away from where we had it. The bug just did that by itself . . . it was totally random."[27] That is exactly the kind of randomness that evolution harnesses so brilliantly.

One of the best ways to demonstrate this would be to *use* evolution. If random mutation is such a powerful source of creative variation, why couldn't we design things that way? We can. The trick is to mimic the way in which living organisms execute those three steps: First, generate variation in a code. Next, find a way for that variation to express itself by performing a task or producing a measurable property. This is essentially what a gene does when it encodes a protein. And finally, use selection to identify the variants that do the very best job of performing that task. If the mechanism is real, we ought to be able to move quickly from random codes to functional and effective ones.

This exact technique has now become a powerful tool in molecular design. In 1994, Willem Stemmer of the biotech firm Affymax sought to "evolve" a new strain of bacterium resistant to an antibiotic known as cefotaxime.[28] He started with a copy of the gene for B-lactamase, an enzyme that breaks down some antibiotics but is not much use against cefotaxime. He mutated the gene (randomly), selected for resistance against cefotaxime in bacteria carrying the mutated genes (there was a little), and then did something very clever. He chopped a copy of the moderately successful mutant gene into small pieces, then allowed them to combine randomly into new sequences that were reinserted into new cells. This randomized swapping of bits and pieces of genes is remarkably close to the kind of gene shuffling that takes place during sexual reproduction, and it was just as effective. In just three rounds of shuffling and selection, he produced mutant proteins that were 32,000 times as effective against cefotaxime as the original protein had been.

The key to Stemmer's dramatic success, as he noted, was his ability to mimic an underappreciated part of the evolutionary mechanism—its ability to shuffle and recombine genes and gene sequences during sexual reproduction. The mechanism of mutation is far more dynamic than is generally believed, and his dramatic demonstration of rapid evolution gave a hint of just how creative mutation can be.

This is but one example of how specific molecules can be designed by harnessing evolution. In 1998, Adam Roth and Ronald Breaker of Yale[29] used similar techniques to produce a powerful enzyme in the laboratory that was unlike anything in nature. Using random DNA sequences attached to the amino acid histidine, they employed eleven rounds of mutation and selection to find a powerful and efficient DNA enzyme that would cut RNA. No DNA enzymes are known in nature, and therefore no biochemist or molecular biologist would have been capable of designing such an enzyme from scratch. With the tools of evolution carefully put to work, they had their RNA-cutter in just a couple of days.

The evolution-in-a-test-tube techniques used by Stemmer, Roth, and Breaker have been used by scores of laboratories around the world, and there is no doubting their effectiveness. In fact, a major movement in computer science is now trying to design "Darwin chips" that will use these very same principles to "grow" circuitry that will solve analytical problems with greater efficiency and speed than human-designed circuitry. Evolution *works,* and it emerges as a powerful creative force that shapes the natural world, and can even be put to work in the service of human industry.

Leaving the Garden

At the beginning of this chapter, I made it clear that religious belief does not require the detection of flaws or inadequacies in evolution. This may not be a radical idea, but it is a far from common one. Indeed, more than once, when a religious acquaintance has discovered that I am a biologist and found it necessary to say something about the great, looming shadow of Darwinism, they have chosen what they obviously hoped would be a diplomatic line of retreat: "Well, you probably realize better than anyone that evolution is *just* a theory. Right?"

Evolution isn't just a theory. We actually use the English term "evolution" in two different ways, and it might not be a bad idea if distinctly different words were coined for those two meanings—history and mechanism—so we'd always get them exactly right.

The first meaning of evolution is history, specifically a living natural history in which the roots of the present are found in the past. The process of evolution, in this sense, describes a natural history shaped by descent with modification. It means that the past was characterized by a process in which present-day species can be traced back to similar, but distinctly different, ancestors. And it means that as we move backwards in time, as we pick up more of the bits and pieces of that historical record, we find a diversity of life that is increasingly different from the life we know and see today. This is, of course, an absolutely accurate description of what we know of life's past on this planet.

In this respect, evolution is as much a fact as anything we know in science. It is a fact that we humans did not appear suddenly on this planet, *de novo* creations without ancestors, and it is a fact that the threads of ancestry are clear for us and for hundreds of other species and groups. It's true that the historical record is incomplete, subject to interpretation and open to revision, especially in the light of new discoveries. We cannot be sure what forms these discoveries might take, and we cannot be sure which of the many threads from present to past have errors or misinterpretations that will some day be corrected. On the issue of whether such threads exist at all, we can be definitive. They do, and that's the point. Evolution is a fact.

What about the second meaning of evolution, as a theory? Darwin's great contribution, as I have emphasized, was not the recognition of evolution as a historical process. Rather, it was his description of a mechanism that could drive evolutionary change. Evolutionary theory is a set of explanations that seeks to account for how that change happened. Evolutionary theory weighs the relative contributions of mutation, variation, and natural seclection, and tries to understand how the interlocking actions of heredity, sex, chance, environment, cooperation, and competition drive the fine details of descent with modification.

Evolutionary theory is a vigorous and contentious field, just as a healthy science should be. Scientific meetings on the subject are filled with argument and disagreement, and that's a good thing. Intellectual conflict, even on a personal level, is good for science because it motivates individual scientists to test their ideas and those of their competitors in the crucibles of experiment and observation. In this respect, evolution meaning the detailed mechanism of change is *theory,* but theory in this context does not mean a haphazard guess or a hunch. Evolutionary theory is not a guess about the nature of life any more than atomic theory is a guess about the nature of matter, or germ theory is pure speculation on the nature of disease. Evolutionary theory is a well-defined, consistent, and productive set of explanations for how evolutionary change takes place.

Evolution is *both* a fact and a theory. It is a fact that evolutionary change took place. And evolution is also a theory that seeks to explain the detailed mechanism behind that change.

It would be nice to pretend, as many of my scientific friends do, that the study of evolution can be carried out without having any effect on religion. In their own way, they might envy other scientific fields—say, organic chemistry or oceanography—that seem to barrel ahead at full speed without ever being cast into the arena to grapple with the Almighty. However one might hope that to be the case, and much of the scientific establishment surely wishes it were, the clash between evolution and religion is not about to go away anytime soon.

The heart of the matter is that evolution is, by definition, a story of origins. This means, however powerful its scientific support, it really does supersede another creation story—in particular, the creation story at the very core of the Judeo-Christian narrative. The conflict between these two versions of our history is real, and I do not doubt for a second that it needs to be addressed. What I do not believe is that the conflict is unresolvable.

The imagination of an age is shaped by its experiences. And these in turn are colored by its knowledge and understanding of the natural world. For thousands of years, humankind thought itself the focus of the universe, watching the sun and moon revolve around a human center, and picturing the stars as decoration on a distant ceiling of the nighttime sky. This view was confirmed by a story of origins detailing our species and all others brought abruptly into existence by the direct action of the Creator. The great legacy of the scientific age is the understanding that neither of these stories is scientifically correct.

Religious people have long since adjusted their views of the celestial universe to include the realization that we inhabit a small planet that moves around a smaller-than-average star, occupying a peripheral position on a nondescript galaxy in the vastness of space. In retrospect it's difficult even to imagine why this view of the cosmos should ever have been viewed as a threat to religion, but so it was. How can we understand this historic hostility to something that we now regard as commonplace?

The answer is not found in religion itself, but in the customs, traditions, and even the social views of the age. By any definition, the idea of God is something that surpasses human understanding. Each age does its best to construct an image of His presence in the universe according to words and images that fit the imagination of the times. For a dozen centuries, the medieval view of God as Lord of the great manor—with his home literally in the sky—was particularly easy for the people of Europe to grasp. As teachers come to appreciate, it is perilously easy to confuse metaphor with reality, to forget that the image we fix in our minds to help understand a difficult concept is not the concept itself.[30]

When this view of the heavens was disturbed by the new technology of astronomy, it was all too easy to take the observations of Copernicus and Galileo as threats to the very idea of God. Four centuries later, theologians find themselves regarding a greater and grander God whose domain was enhanced, not diminished by the discoveries of science.

The concept of a God whose prodigious power was concentrated into a few spectacular days of ecosystem-building was easy to grasp. The view of God as maker and master of Eden has an enduring appeal, and served well to crystallize the relationship between Creator and created, especially for the first forty-eight of the fifty centuries following the Book of Genesis. The personal and physical presence of God in the Eden of Genesis was a source of comfort to those who sought to understand the complexity and the beauty of the living world, the diversity of which remained a mystery to even the most careful observer for most of human history.

In the last two centuries, things have changed. Biology has developed from a purely descriptive science into a constructive one, and we now understand the genuine source of life's diversity. We also have lengthened our view to see that the story of life includes a grand, even heroic past, a record of change and struggle, of failure and triumph. As we add to the growing richness of life's documentary record, we can be justifiably proud, not just of the fact that we—along with every other living thing on the planet—are among life's winners, but especially of the fact that we are the very first creatures in 35 million centuries to become aware of the magnificence of our legacy.

It is high time that we grew up and left the Garden. We are indeed Eden's children, yet it is time to place Genesis alongside the geocentric myth in the basket of stories that once, in a world of intellectual naivete, made helpful sense. As we walk through the gates, aware of the dazzling richness of the genuine biological world, there might even be a smile on the Creator's face—that at long last His creatures have learned enough to understand His world as it truly is.

3

GOD THE CHARLATAN

When God and science come together, common wisdom is that something has to give. For most people, that something is not the whole of science, but rather the troubling subset of biology called evolution. The power and utility of modern science is far too appealing for most religious people to discard it altogether, but evolution is different. It claims to have discovered a material, mechanistic *reason* for our existence. It says that we are here as the result of natural processes, the very same natural processes that we see operating in the world around us all the time. And there's the problem.

According to one line of thinking, unless we show that purely natural processes are incapable of producing living things like us, we have excluded the Creator from any effective role in His creation. So, we'd better find a way to explain why those natural processes are not sufficient.

At least in part, this viewpoint may come from our particular conception of the natural world. Western religions do not attribute intelligence, purpose, or intention to the wind or the seas or the snow. In our intellectual tradition, to see such consciousness in nature would be manifestly pagan and maybe even anti-Christian. The Western vision of the material world is distinctly impersonal and produces a profound sense of division between man and nature.

Traditionally we look to an intelligence that transcends nature to give life meaning. That intelligence is God. A perfect example of this viewpoint was articulated by author and political commentator William F. Buckley in his opening statement at a debate on evolution held in December of 1997 on his Public Broadcasting television show, *Firing Line*:

> How much science do we need to master to qualify as reasonably to affirm that there has to be a reason for you and me and the world we live in—a reason other than raw nature driven by—driven by what?[1]

To Buckley, and to many others, "raw nature," which is exactly what evolution represents, just cannot be enough. Something is needed to *drive* nature, to provide the direction and meaning that we sense in life. Stating it simply, we'd better find something *else* out there, or we're in trouble.

If I had any doubts about the allure of this viewpoint, they came to an end many years ago in a brief conversation at the entry desk of my university's library. At the urging of students, I had just agreed to enter a debate against Henry Morris, the founder and president of a California-based group known as the Institute for Creation Research (ICR). Actions of the legislatures of several states had placed the so-called evolution-creation controversy into the national news. The upcoming debate on our campus was being vigorously publicized with posters, one of which was taped to the desk right at the entrance to the library.

I had been at Brown for only a few months, but was already a familiar face at the science library. The attendant gestured at the poster to ask if I was going. "You bet," I said with some enthusiasm, not letting on that I was going to be one of the principals. "Well," he said, drawing back in his chair to affect an air of wisdom earned for being many years my senior, "I don't know about this evolution stuff. But I do like to think that we were put here for a reason, don't you?" How could I possibly argue with that? I gave him a nod and went on my way, preparing to do my best to demolish the very idea that he held so close.

His prime objection to evolution, as he had made perfectly clear, did not derive from a careful analysis of the scientific evidence, pro and con. It came instead from a deep emotional commitment to the idea that *nature alone* could not be the source of *human nature*. At its core, evolution threatens the sense of specialness we enjoy in a world where we have come to view ourselves as the centerpiece of creation. Its opposite number, opposite creation, is an attractive, appealing, and powerful view. It is also demonstrably, completely, and even tragically wrong.

All the Wrong Places

If we accept, a priori, the view that natural processes *cannot* be sufficient to account for our presence in the world, we've got some serious decision-making to do.

We could, for example, allow that evolution might have produced every species except for us. Even the staunchest opponents of evolution do not take this position, and for very good reason. We humans are living creatures and share nearly every aspect of our biological existence with other living things. A human muscle, nerve, or bone cell is not particularly different from a cell taken from the same tissue of another mammal. Our genetic instructions are encoded in the same language of DNA, and human genes transplanted into other animals, bacteria, and even plants function perfectly. These sweeping similarities would make it silly to pretend that evolution could produce any species except for *Homo sapiens*. There just isn't enough about us that is biologically different from other animals to say that evolution applies to them, but not to us.

We have to find another place to draw the line.

We could, if we were especially cautious, draw that line in a way that *includes* as much science as possible. We might, for example, accept the general picture that historical geology has given us for the age of the earth. That would put us at peace with the physical sciences. We might further agree to the general validity of the fossil record and its sweeping pattern of descent with modification. That would keep the paleontologists off our backs, and save us from repeated attack as new discoveries flesh out further detail in evolutionary history. But we would still have to find at least one essential event in the history of life to stand outside these natural processes—one thing that *must* have been done by the Creator. And ideally, we'd make this an event so tiny and so distant in the past history of life that no historical record of how it actually occurred could ever be found. We could, following this strategy, argue that evolution cannot account for the biochemical machinery of the living cell. That's where we could claim that a designer is required, and that's how we would protect our worldview against the ravages of evolution. Such is the viewpoint espoused by Michael Behe and others who hold up the lofty banner of "intelligent design." It's a viewpoint we will dissect in Chapter 5, "God the Mechanic."

If we were a bit bolder, we might draw the line to include a little less of science. We could take a stance that would neutralize the historical record of change that is at the heart of an evolutionary account of life's history. If we were successful on this score, we would deprive evolution of one of its greatest strengths—the fossil record. That record is a powerful tool in the hands of evolutionists, because it documents how evolution actually took place. If we were able to establish that it shows no such thing, we just might kick the foundations out from under Darwin's temple. We could do this by showing that natural processes cannot produce evolutionary change, and therefore the pattern of succession so apparent in the record must come from somewhere or *someone* else. And you know who that someone might be! This is the strategy taken by Phillip Johnson and his associates, and we will examine its implications in Chapter 4, "God the Magician."

Or we might adopt a more direct and more spectacular strategy. We could reject *everything* that appears to support evolution in even the slightest respect. That strategy is the topic of this chapter.

Such is the road traveled by my debate opponent of many years ago, Henry Morris, and his colleagues at the ICR. They have plenty of company, of course, in other institutes, organizations, and publishers around the world. Of all the attacks against evolution, theirs is the boldest and most consistent, rejecting out of hand anything and everything that seems to support evolutionary biology. Among the anti-evolution crowd, the ICR and their allies are known as "YECs"—young-earth creationists—a name that indicates how thoroughly they have departed from the scientific mainstream. They believe that the earth is young—no older than 10,000 years—and they are prepared to defend that belief.

Why is it so important to believe in a young earth? Fitting the chronology described in a literal reading of the Bible[2] is only part of the reason. The larger goal is to invalidate the mountains of historical evidence that evolution took place. Therein lies the appeal of this line of attack, for if the earth really is only a few thousand years old, then the fossil record is a meaningless illusion.

To say the least, there are problems with this view. As I have emphasized, what impresses one most about the fossil record is its sequential character. Why, you might ask, do the geological ages contain so many *apparent* evolutionary sequences if evolution never took place? The YECs have an answer. It was all caused—accidentally—by the Flood of Noah. This is the theme of *Genesis Flood*,[3] a 1961 book by Morris and John Whitcomb that remains a classic among young-earth creationists.

The book's primary thesis is that the *appearance* of evolution in the fossil record is an accidental artifact of sedimentation. When a great flood destroyed most of the life on planet earth, say the YECs, those living things that could not escape to higher ground wound up near the bottom of the sediment. These are the rocks that earth scientists regard as being from the oldest geological ages. The young-earth creationists contend that those animals that could swim or float or climb most effectively wound up near the top, in so-called recent ages. That is their counterexplanation for science's finding that human fossils do not appear until the last few million years of the geological record. Without exception, they hold, the humans alive at the time of the Noah's Flood survived until the last few meters of sedimentation. Thirty-nine days of treading water, you might say.

It doesn't take an expert in historical geology to appreciate that there is something wrong with this explanation. The actual pattern of life over geologic time doesn't even come close to matching what would have happened in a single worldwide flood.

Flowering plants, for example, are recent appearances on this planet (see Figure 2.1 in Chapter 2). Although our landscapes are dominated by these organisms, from the Amazon rain forest to the Great Plains, from the European steppes to the central African jungle, not a single fossil of an Anthophyte (the scientific name for such plants) appears in the first 2 billion years of the geological record. Plants are good at many things, but running to higher ground during a flood is not one of them. The great coal forests of the Carboniferous periods contain exquisitely preserved fossils of club mosses, giant ferns, and horsetails. They do not include so much as a single flower. Not a dandelion, not a rose, not an acorn, not so much as a mustard seed. The turbulence of a flood such as Noah's would have churned all plant material together, cementing seed and flower and broken stem into a single layer, testament to simultaneous existence. For flowering plants, no such testament exists. Why not?

The answer, as geologists have known for more than a hundred years, is simple. At the time of the Carboniferous, flowering plants had not yet *evolved*.

Is there a chance that these "flood geologists" are genuine, sincere scientists? Is it possible that they are lonely pioneers laboring in a great and noble tradition of scientific outcasts, fighting for respectability and ultimately for proof of their ideas? I don't think so; and I say that not to make a character judgment, but as an evaluation of scientific behavior. If they really believe in the validity of their interpretations of fossil history, they should be charging ahead to exploit a great scientific opportunity: the coprolite.

Coprolites are fossilized feces. Over the years, paleontologists have found thousands of these objects, including a notable one described in a paper from *Nature* magazine in 1998[4] with the alluring title, "A King-sized Theropod Coprolite." The size and location of the object indicate it was produced by a meat-eating dinosaur, probably *Tyrannosaurus rex*. The coprolite was packed with large fragments of partially digested bone. Other coprolites are available, if our YEC colleagues prefer, from ancient mammals, plesiosaurs (swimming reptiles), and even from insects.

For young-earth creationists, these remains present a stunning opportunity to validate their ideas. All they would have to do is pick through these objects and find evidence of a single contemporary organism. Seeds or microscopic pollen grains from modern plants would do the trick in the case of herbivorous dinosaurs. If they could just find a couple of tuna bones in the stomachs of those plesiosaurs, they'd stand the geological world on its head by demonstrating that creatures of the "ancient" and the modern worlds existed side by side before the flood, as they have always maintained. The young-earth creationists make no such effort. They keep themselves carefully aloof from any hands-on contact with genuine evidence, such as the fact that the digestive systems of the plesiosaurs are filled with ammonites, extinct mollusks from the same geological age.

Further demonstrations of the hopelessness of flood geology are not necessary here, although several detailed examinations of this nonsense have appeared in print.[5]

In this and the next two chapters I will examine the three anti-evolutionary strategies touched upon above. My plan is to go right to the basic flaws of each argument, one at a time, and demonstrate how spectacularly inconsistent each is with the actual scientific evidence.

These refutations will, when we are finished, beg several much larger questions. Why do these badly flawed attacks on evolution persist? Why do they have so much appeal? Why (with apologies to country singer Johnnie Lee) are these folks *looking for God in all the wrong places?*

I believe it has very little to do with the science of evolution, and everything to do with how that science is misapplied to the larger questions of human existence. We'll get to those questions, and eventually we'll see how neatly evolution folds into the fabric of scientific reality and spiritual belief. But first, let's follow one line of argument from these anti-evolutionists and see where it leads.

How Old Is the Earth, Really?

It's a fair question. I first approached the creationist literature with only the most elementary knowledge of geology and with the unspoken trust that a scientist in one field has in the work of another. I trusted that geology, like biology, was an active and vigorous science, that its practitioners would have challenged any fundamental conclusions that were on shaky ground, and therefore that the broad, general statements of geologic history could be relied upon. One of those statements was that the earth is several billion years old.

The very first time I read through the creationist literature on this subject, a sense of excitement came over me. Most scientists, myself included, are not interested in careers devoted to the confirmation of orthodoxy. What could be duller than publishing, year after year, that current theories on one topic or another are correct? All genuine scientists dream of being revolutionaries. They dream of striking out against the prevailing view, having the imagination to contradict a scientific establishment, and then gathering the experimental evidence to show that they are right. The thought that there might be just such a revolution going on right here was intriguing.

I dug into Henry Morris's *Scientific Creationism,* one of the best-known books on the topic, expecting to find a genuine scientist making a consistent case out of evidence for a young earth. Instead, I found a curious ambivalence to commit to *any* actual age for the planet:

> As a matter of fact, the creation model does not, in its basic form, require a short time scale. It merely assumes a period of special creation sometime in the past, without necessarily stating when that was.[6]

Fair enough, I supposed. It's a useful debating strategy to make yourself appear to be open-minded while depicting the other guy as wearing the intellectual straitjacket of needing the "great ages" theory to support evolution. But where were the scientific revolutionaries? Where were the crusaders who thought that the establishment was off by a factor of 400,000? Where was their stunning, irrefutable evidence? In another book, Morris approached the problem by admitting that some dating methods used by geologists gave very old ages for rocks and minerals, but that

> other processes do give much younger ages, however. For example the present rate of sedimentary erosion would have reduced the continents to sea level in 6 million years and would have accumulated the entire mass of ocean-bottom sediments in 25 million years. Present rates of volcanic emissions would have produced all the water of the oceans in 340 million years and the entire crust of the earth in 45 million years.[7]

The self-contradictions of time and logic in this passage left me scratching my head. Even my middle-school earth science class had prepared me to realize that processes like volcanism and erosion are *opposites*. The land produced by one is balanced by the land eroded away by the other, and therefore neither could provide a useful age for the planet as a whole.

Sadly, nearly all of the alternate methods proposed to date the earth were likewise flawed. Morris noted that the amount of certain minerals found in seawater could have been produced in just a few thousand years given the present rate of erosion. The amount of silicon found in all the world's oceans, for example, could be accounted for by just 8,000 years of river inflow. That sounds impressive until one reads further and discovers that the amount of aluminum dissolved in seawater could have been produced in just 100 years!

I think even Henry Morris might be willing to admit that the earth is older than 100 years. What these values actually measure, of course, are the residence times of various materials in seawater. Each year, the rivers of the world dump into the ocean an amount of dissolved aluminum compound equal to roughly 1/100th of the oceans' existing aluminum content. Aluminum is a very reactive element, and readily precipitates with other compounds to form kaolinite, one of the main chemical constituents of clay. Silicon and other minerals undergo similar reactions that take them out of solution at rates high enough to make their concentrations useless as tools to measure the age of the planet.

For a moment, I thought I had found something. Morris pointed out that the strength of the earth's magnetic field has been steadily decaying over the 150 years or so that we have been able to measure it. If we extrapolate this decay backwards in time, then as little as 10,000 years ago the field would have had to have been unimaginably strong. Meaning? That the earth was less than 10,000 years old.

I wondered about the conspiracy of scientific silence that had kept this crucial fact hidden away during my education. I was positive that this measurable, provable physical fact was the key to a young earth. I thought about it for a while, checked the references on magnetic field measurement, and decided I was ready to confront an expert. I called up Peter Gromet, a well-known geologist on campus, and asked if I could come by. Later that afternoon I prepared. I gathered up my charts and tables, summarized my arguments in graphic form, and made certain that I was ready. Well prepared, as it turned out, to embarass myself.

Peter listened patiently, and broke into a smile, then a gentle laugh. "You've never had a course in geology, have you?" I hadn't, of course. As Gromet patiently explained, when rocks are formed, their magnetic minerals orient themselves in the earth's magnetic field. This means that the rocks of the earth's crust carry within them a record of the strength and direction of that field at the time of their formation, a record known as paleomagnetism. Paleomagnetism allows geologists to trace the strength of the field. From core samples of sedimentary rock an interesting fact has emerged: The earth's magnetic field has gone through innumerable reversals and fluctuations over time. In other words, the current weakening of the field that Morris referred to is nothing more than the latest ebb in a series of fluctuations that come and go with surprising regularity.

Using the 150-year decrease in the field's strength to argue for an earth of 10,000 years makes about as much sense as watching the tide rise by a foot over three hours and concluding that the earth could be no more than nine years old![8] It just ain't so, and their use of the magnetic field strength argument destroyed any hope I had that the creationists might have been on to something.[9]

So, how old is the earth, really?

This was an issue that troubled Darwin, who realized that his theory was absolutely dependent upon a long history of life for natural selection to have worked its magic. He resisted any attempts to fix the age of the planet at less than 100 million years, and clearly understood that a much younger age for the earth might well deal a serious blow to evolution. There simply wouldn't have been enough time for it to work.

Against this backdrop came the calculations of William Thomson, Lord Kelvin, one of the deans of classical physics. Kelvin calculated the length of time required for a body the size of the earth to cool from an initial molten state to its present temperature, a technique that fixed the age of the planet between 20 million and 400 million years. He was later to refine these calculations to 98 million years, and in 1893 endorsed a further refinement that found 24 million years as the most probable age for the planet.

There is no doubting the discomforting effect these calculations had on Darwin. Today we can look back on Kelvin's calculations as no more than an amusing example of very bad timing. In 1896, A. Henri Becquerel discovered that uranium salts emitted invisible rays that were capable of exposing photographic film. Radioactivity, as this phenomenon was to be known, supplies a persistent source of heat deep within the earth, invalidating Kelvin's careful calculations. It also provided something else—an accurate way to fix the true age of the planet.

A Radioactive Stopwatch

Uranium, or ^{235}U, the well-known radioactive isotope, decays through a series of intermediates to an isotope of lead known as ^{207}Pb. The half-life of the series is 713 million years, which means that after that length of time, one half of the ^{235}U in a sample will have decayed into ^{207}Pb. Despite this decay process, the intensity of uranium mining tells us that there is still plenty of ^{235}U available on the earth's crust. And this fact leads to a bold and remarkable conclusion—the earth could not have existed forever! The planet really did have a beginning.

Since the rate of decay of ^{235}U is constant, if this planet had existed forever, by now *all* of the ^{235}U from the planet's formation would have vanished. That is not the case, of course, and therefore our planet's beginnings are not in the infinite past. Can we go a little further and make some estimates as to just how old the planet really is?

Absolutely. To be sure of the age of a rock, all we have to do is to learn the amount of ^{207}Pb that was present when the rock was formed, and then to determine the amount of new ^{207}Pb that has been added by radioactive decay. The initial ^{207}Pb is known as primordial lead, and the new ^{207}Pb is called radiogenic lead. Unfortunately, primordial and radiogenic lead are chemically identical, so there is no way to tell them apart. That might seem to be a serious problem, but in fact it is easily surmounted.

Another of the isotopes of lead, ^{204}Pb, is nonradiogenic, which means that it is never formed by any decay process—*all* of the ^{204}Pb in any rock is primordial. When a rock is formed, since ^{204}Pb and ^{207}Pb are chemically identical, it will incorporate the two isotopes at the ratio at which they are found on the surface of the earth at that time. Then, once the formation is complete, ^{235}U will begin to decay into ^{207}Pb. If a rock is young, its $^{204}Pb/^{207}Pb$ will be nearly identical to the current ratio of these two isotopes. If billions of years have passed, however, a geologist will notice two things about the mineral: one, it contains very little ^{235}U; and two, the $^{204}Pb/^{207}Pb$ will be very low, because of the accumulation of ^{207}Pb over time. With some care, the actual amount of ^{207}Pb accumulation can be measured, and the age of formation of the rock can be calculated.

There are three independent uranium and thorium methods for dating rock, each based on a different isotopic series, and each providing an independent check upon the others. The decay of radioisotopes of ^{147}Sm (samarium), ^{176}Lu (lutetium), and ^{187}Re (rhenium) are also used to determine the ages of rocks and minerals, each presenting its own advantages and each providing an independent way to check ages determined by other methods.

One additional method worth mentioning is the potassium-argon technique. Potassium-argon dating takes advantage of the decay of one of the radioisotopes of a common element, potassium (^{40}K), into a rare gas, argon (^{40}Ar). Geologists often choose a crystalline mineral whose chemistry requires a specific number of potassium atoms at fixed positions in the crystal, but contains no initial argon. As time passes, more and more ^{40}K is transfomed into ^{40}Ar, which is then trapped within the lattice of the crystal. A newly formed volcanic rock will have almost no argon, which is rare in the atmosphere. But an ancient rock will show a loss of ^{40}K exactly balanced by a gain of ^{40}Ar, enabling geochemists to determine its age based on the half-life of ^{40}K.

The possible errors from using this method illustrate the caution that professional geologists apply to their work. The potassium-argon method is popular with geologists because the only serious source of error arises when a rock has been disturbed by heating, allowing the accumulated argon gas to escape. In all such cases, the time span determined is an underestimate of the rock's true age, never an overestimate.

If this planet were recently formed, then rocks and minerals everywhere would show lead isotope ratios and uranium/lead ratios that match the current abundances of these materials on the surface of the earth. Potassium-containing minerals on a recently created earth, just like newly formed rocks from active volcanoes, would contain little or no argon. Think about that for a moment: The presence of radioactive isotopes in rocks could *prove* that the earth is not infinitely old, and furthermore could provide us with the chance to determine if the earth is really as young as the creationists claim it is. What an opportunity! Darwinism could have been smashed against the anvil of nuclear physics.

Remember that geologists, following the work of Smith and the other pioneers, had established a series of well-defined geological ages. To Darwin and anyone else who regarded these ages as genuine, they represented successive chapters in the living history of our planet. To creationist critics of evolution, including the authors of *Genesis Flood,* the geological ages are illusory. To them, the rocks were formed more or less simultaneously, and only in the prejudiced imaginations of evolutionists do they fit into a temporal sequence. For creationists, the unexpected discovery of radioactivity should have been a godsend. Imagine—the chance to destroy once and for all the dogma, as they saw it, of geological ages, and to establish the truth of simultaneous creation.

One might have assumed that the opponents of evolution would have embraced the definitive results of radiometric geology. One can imagine whole laboratories of young creationist geologists using their powerful new tools to overturn an entrenched scientific establishment, and build their own careers to public and professional acclaim.

No such luck. The reason? The ratios that emerge, time and time again, indicate that the oldest rocks on the earth approach an age of 4.5 billion years. In one study after another, these techniques confirmed and refined the traditional system of geological ages. Not only had the geologists of the nineteenth century gotten the sequences right, but they had been much too conservative in their estimates of the duration of those ages. These stunning results confirmed, long after his death, exactly what Darwin had suspected—the age of the earth was much, much greater than the critics of evolution had argued.

The discovery of radioactivity forced all of Darwinism to undergo an unexpected test at the lab benches of the physical sciences. Literally everything was on the line, as radiometric dating made it possible to test every assumption in the timescale of evolution. What happened? Evolution passed, and it passed with flying colors. Indeed, Darwin would have been happy with physical confirmation of a living history of just 100 million years; in fact, he got more than ten times as much.

A Census of the Universe

The power of radioisotope methods is so great that creationists are reduced to sniping about the chemical and physical assumptions inherent in the techniques, to suggesting that decay rates are variable (some are, but by such tiny amounts that the effect on dating is insignificant), or to suggesting, as we will see at the conclusion of this chapter, that the results just don't matter.

Is it possible that geologists have somehow biased their experimental results? Perhaps they have selected *only* those minerals and isotope series that support their preordained conclusions. Is there a way we can be sure that we have got things right? Well, there are at least two.

If we are worried that scientists tend to select only those samples that fit their preconceptions of the great ages required by evolution, we can get around that problem in the simplest way imaginable. We can select *everything*. There is a substantial literature on the abundance of various elements in the cosmos, drawing upon studies of the earth, the moon, cosmic dust, and meteorites. From this list we can extract information about which of the isotopes of the various elements, known as *nuclides,* are found in nature at detectable levels. Table 3.1[10] shows a list of the known radioactive nuclides with half-lives of 1 million years or more. Of the thirty-four known radioactive nuclides, only twenty-three are found in detectable amounts in nature. List them in the order of their half-lives, and an interesting pattern begins to emerge:

TABLE 3.1: LISTING OF NUCLIDES BY HALF-LIFE

Nuclide	Half-Life (years)	Found in Nature?	Nuclide	Half-Life (years)	Found in Nature?
^{50}V	6.0×10^{15}	yes	^{244}Pu	8.2×10^{7}	yes
^{144}Nd	2.4×10^{15}	yes	^{146}Sm	7.0×10^{7}	no
^{174}Hf	2.0×10^{15}	yes	^{205}Pb	3.0×10^{7}	no
^{192}Pt	$\sim 1.0 \times 10^{15}$	yes	^{236}U	2.39×10^{7}	yes—P
^{115}In	6.0×10^{14}	yes	^{129}I	1.7×10^{7}	yes—P
^{152}Gd	1.1×10^{15}	yes	^{247}Cm	1.6×10^{7}	no
^{123}Te	1.2×10^{13}	yes	^{182}Hf	9×10^{6}	no
^{190}Pt	6.9×10^{11}	yes	^{107}Pd	$\sim 7 \times 10^{6}$	no
^{138}La	1.12×10^{11}	yes	^{53}Mn	3.7×10^{6}	yes—P
^{147}Sm	1.06×10^{11}	yes	^{135}Cs	3.0×10^{6}	no
^{87}Rb	4.88×10^{11}	yes	^{97}Tc	2.6×10^{6}	no
^{187}Re	4.3×10^{10}	yes	^{237}Np	2.14×10^{6}	yes—P
^{176}Lu	3.5×10^{10}	yes	^{150}Gd	2.1×10^{6}	no
^{232}Th	1.40×10^{10}	yes	^{10}Be	1.6×10^{6}	yes—P
^{238}U	4.47×10^{9}	yes	^{93}Zr	1.5×10^{6}	no
^{40}K	1.25×10^{9}	yes	^{98}Tc	1.5×10^{6}	no
^{235}U	7.04×10^{8}	yes	^{153}Dy	$\sim 1.0 \times 10^{6}$	no

"Yes" indicates that a nuclide is present in detectable amounts in nature. "Yes—P" indicates that a nuclide is present in detectable amounts, but that it is continuously produced as the product of another decay series.

All of the nuclides at the top of the list, where the longest half-lives are located, are found in nature. Most of the nuclides at the bottom of the list, the ones with shorter half-lives, are not. Most, but not all. And that's where things get really interesting.

If we consult our friendly local nuclear physicist, we will discover something revealing about the short-lived nuclides that are found in nature. For example, there is plenty of ^{10}Be (beryllium-10) around, for a very good reason. ^{10}Be is continually produced by cosmic ray bombardment of dust particles in the upper atmosphere, and very small (but detectable) amounts of this nuclide are incorporated in rainfall and sediment. ^{53}Mn (manganese-53) is produced in exactly the same way. ^{237}Np (neptunium-237) is not found on the earth at all, but is produced by cosmic ray bombardment on the moon (remember, *everything* that has been analyzed is included in this list). In fact, if we strike from the list every nuclide[11] that is continually produced by natural processes, we should be left only with those that persist from the date of the formation of our solar system. When we do that, the data fairly shout to us:

TABLE 3.2: LISTING OF *PERSISTENT* NUCLIDES BY HALF-LIFE

Nuclide	Half-Life (years)	Found in Nature?	Nuclide	Half-Life (years)	Found in Nature?
^{50}V	6.0×10^{15}	yes	^{40}K	1.25×10^{9}	yes
^{144}Nd	2.4×10^{15}	yes	^{235}U	7.04×10^{8}	yes
^{174}Hf	2.0×10^{15}	yes	^{244}Pu	8.2×10^{7}	yes
^{192}Pt	$\sim 1.0 \times 10^{15}$	yes	^{146}Sm	7.0×10^{7}	no
^{115}In	6.0×10^{14}	yes	^{205}Pb	3.0×10^{7}	no
^{152}Gd	1.1×10^{15}	yes	^{247}Cm	1.6×10^{7}	no
^{123}Te	1.2×10^{13}	yes	^{182}Hf	9×10^{6}	no
^{190}Pt	6.9×10^{11}	yes	^{107}Pd	$\sim 7 \times 10^{6}$	no
^{138}La	1.12×10^{11}	yes	^{135}Cs	3.0×10^{6}	no
^{147}Sm	1.06×10^{11}	yes	^{97}Tc	2.6×10^{6}	no
^{87}Rb	4.88×10^{11}	yes	^{150}Gd	2.1×10^{6}	no
^{187}Re	4.3×10^{10}	yes	^{93}Zr	1.5×10^{6}	no
^{176}Lu	3.5×10^{10}	yes	^{98}Tc	1.5×10^{6}	no
^{232}Th	1.40×10^{10}	yes	^{153}Dy	$\sim 1.0 \times 10^{6}$	no
^{238}U	4.47×10^{9}	yes			

Every nuclide with a half-life of less than 80 million years is *missing* from our region of the solar system, and every nuclide with a half-life of greater than 80 million years is present. Every single one. These data are an unbiased atomic sampling of our corner of the known universe. And the results are crystal-clear. There is a reason that the short-lived nuclides are no longer around, and the reason is obvious: The solar system is much older than 80 million years. In the billions of years since its formation, the short-lived nuclides have simply decayed themselves out of existence.[12]

Ten or twenty half-lives should be enough to cause most nuclides to disappear from the universe. After one half-life, only one half of the original nuclide would be left; after two half-lives, one quarter. After ten half-lives, only $1/2^{10}$, or less than 0.1 percent, of the original amount of a nuclide will remain. After twenty half-lives, just $1/2^{20}$ will still be found, less than 0.0001 percent.[13] Twenty half-lives of 80 million is 1.6 billion years. And this means, unequivocally, that if the solar system was younger than a billion years or so, at least a few of those short-lived nuclides ought to persist from its original formation to tell us about it. There are none to be found.

Instead, Table 3.2 shows us a line in the sand, drawn eloquently, not by experimental bias, but by the elemental composition of the universe itself. The universe had a definite beginning—that's why we can see the line—and that beginning was several billion years ago. No doubt about it.[14]

The Rock That Sings (Concordancy)

From the general, we can move to the specific. If the composition of the universe itself lets us know that it was fashioned several billion years ago, can we pick up specific rocks and fix their own dates of formation even more precisely? Once again, the answer is yes.

One of the long-lived nuclides in Table 3.2 is rubidium-87 (^{87}Rb), which decays to ^{87}Sr over a half-life of 48.8 billion years. There are also three isotopes of strontium (^{88}Sr, ^{86}Sr, and ^{84}Sr), which are not produced by any radioactive decay process but are chemically indistinguishable from ^{87}Sr. Let's suppose that a mineral is formed that contains both rubidium and strontium. What will happen to the chemical composition of that mineral as time goes by? Over time, while the amounts of nonradiogenic Sr remain constant, the amount of ^{87}Sr will increase by the exact amount that ^{87}Rb decreases. Therefore, in a very old rock, the ratio of ^{87}Sr to ^{86}Sr, a nonradiogenic isotope, will have increased dramatically.

If we knew the starting ratio of rubidium to strontium, we could calculate the exact age of the mineral. But how can we possibly know the starting conditions of a rock formed millions or even billions of years ago? Here's where the rubidium-strontium method sets itself apart—it provides a method to do just that.

Let's take a typical rock, which is composed of several different minerals. Some of those minerals will have lots of rubidium but little strontium (a high Rb/Sr ratio), others will have very little rubidium but plenty of strontium (a low Rb/Sr ratio), and still others will have intermediate amounts. At the moment the rock is formed, we could represent the ratios of various isotopes as in Figure 3.1:

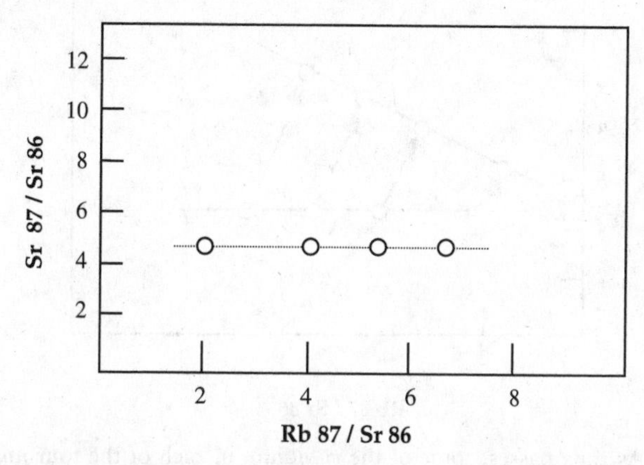

Figure 3.1. Rubidium and strontium isotope ratios as they might appear in a newly formed rock. Four chemically different minerals (open circles) are found in the rock. Because all strontium isotopes are chemically identical, ^{87}Sr and ^{86}Sr are incorporated into each of the minerals at the same ratio (around 4.8). However, each mineral incorporates a *different* amount of rubidium and strontium, giving each one a different ratio of ^{87}Rb to ^{86}Sr (values shown here range from 2.0 to 6.7).

Because the four minerals shown in this diagram are chemically different from one another, they take up Rb and Sr in different proportions, leading to ^{87}Rb/^{86}Sr ratios that vary among them. However, the ^{87}Sr/^{86}Sr ratio is identical in each mineral, which also makes sense—remember, these two isotopes are chemically identical, and therefore there is no way for any mineral to include one of the isotopes preferentially over the other. That's why a graph of the isotopic ratios of the four minerals in our hypothetical rock appears as a flat line. Each mineral starts with an identical ratio of the two strontium isotopes, but a different ratio of rubidium to strontium.

What will happen as time passes? As rubidium decays to strontium, the amount of ^{87}Rb in each mineral will decrease, and the amount of ^{87}Sr will increase. But remember that the amount of that increase is directly proportional to the amount of ^{87}Rb in the mineral. So, those minerals with lots of rubidium will accumulate a great deal of ^{87}Sr over time, and those with only a little rubidium will accumulate much less. What will this look like? (See Figure 3.2):

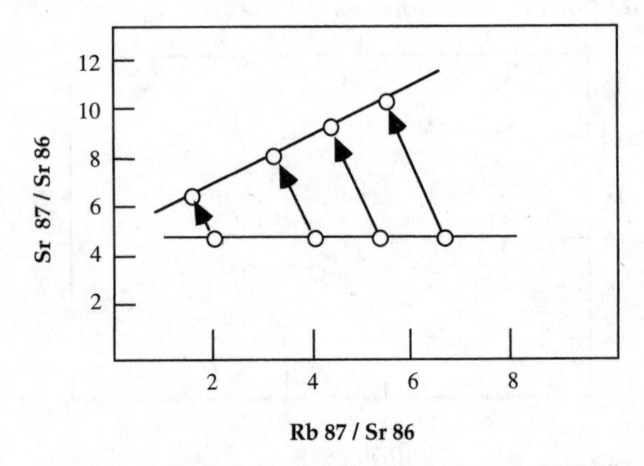

Figure 3.2. As time passes, some of the rubidium in each of the four minerals decays to produce strontium. This decreases the ^{87}Rb/^{86}Sr as the rubidium is lost but increases the ^{87}Sr/^{86}Sr as ^{87}Sr accumulates. If the rock is undisturbed, the values for the four minerals will form a straight line, an isochron whose slope determines the age of the rock.

As time passes, the ^{87}Sr/^{86}Sr ratio will change in each mineral, but in every case it changes in direct proportion to the rubidium/strontium ratio in the mineral when it was formed. As a result, our points will still lie on a straight line, and the slope of that line gives us a measure of the amount of time that has passed since the formation of the rock. We do not need to make an estimate of the starting conditions, because the starting conditions can be determined directly.

The power of this method is remarkable. Every single mineral in the rock lies on the line, which is known as an isochron (a chart line signifying events that occur at the same time); and therefore every mineral "agrees" on the age of the rock. Each of the many minerals of a complex rock provides a completely independent check upon its age. When

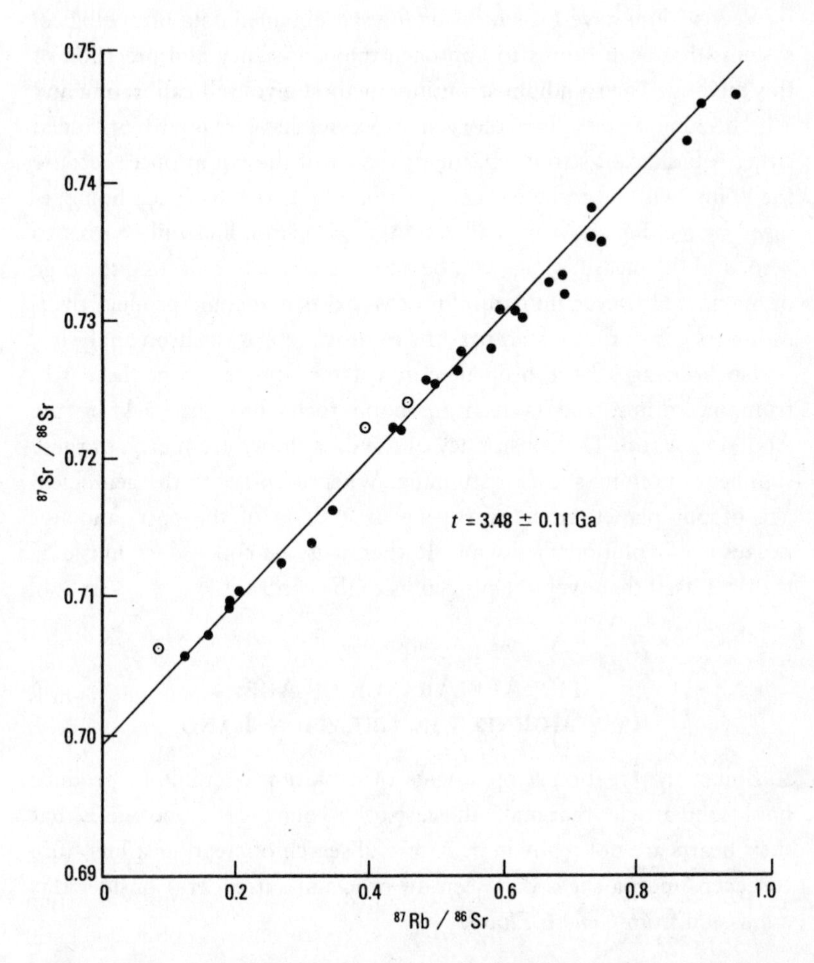

Figure 3.3. The great power of the rubidium-strontium method is that measurements of several minerals within the sample provides an internal test of the starting conditions of the rock. This isochron, showing a sample of the Morton gneiss (named for the nearby town of Morton, Michigan), shows an age of formation of 3.48 ± 0.11 billion years before present.

Figure from G. Brent Dalrymple, *The Age of the Earth* (Stanford, CA: Stanford University Press, 1991).

they fall into such an isochron, the rock is said to be *concordant,* literally "singing together." The concordant minerals of the rock analyzed in Figure 3.3 show a real-world application of the rubidium-strontium method to geology.

Very seldom have I (or most biologists) obtained data on biological systems that even begins to approach the consistency and precision of this method. The rubidium-strontium method gives self-calibrating and self-checking results. If geological processes have removed or added either rubidium or strontium, the method will show it at once, because the points will fail to lie on a straight line. If a rock has been homogenized by melting and recrystallization, the isochron line will be reset to zero, and the measured age will be an underestimate reflecting the time of melting. However, no natural process exists that could produce overestimates of age that would pass the rigorous test of isochron analysis.

Isochron ages have been determined for samples from the earth, from meteorites, and even from moon rocks brought back by the Apollo program. The consistency of the data drawn from each of these samples is nothing short of stunning. When it comes to the geological age of our planet, true controversy is a thing of the past, and not because of evolutionary dogma. Rather, it is the concordant music of the data itself that overwhelms claims to the contrary.

The Appearance of Age: It's Morning in Creation-Land

Although the creationist opponents of evolution continue to produce books and articles that make the case for a young earth, one senses that their hearts are not really in it. A careful search of creationist literature can even find passages that seem to proclaim surrender. Consider this admission from *Genesis Flood*:

> There are many cases now known where the age estimate has been checked by two or more different methods, independently. It would seem improbable that the elements concerned would each have been altered in such a way as to continue to give equal ages; therefore such agreement between independent measurements would seem to be strong evidence that alteration has not occurred and that the indicated age is therefore valid.[15]

Geologist G. Brent Dalrymple, author of *The Age of the Earth,* testified against the young-earth theories of the creationists at the well-publicized Arkansas creation science trial in 1981, and his thorough demolition of their arguments is a matter of public record. As Dalrymple put it most succinctly, "The creationists' 'scientific' arguments for a young earth are absurd."[16]

Case closed. Let us not declare victory on this point prematurely, however, because the creationists have one more trick up their sleeves. Just when they seem about to be overwhelmed by the scientific data on the age of the earth, Whitcomb and Morris, the authors of *The Genesis Flood,* spin 180 degrees in a flash of illogic, and say that a 4.5-billion-year-old earth was *just* what they expected:

> We reply, however, that the Biblical outline of earth history, with the geologic framework provided thereby, would lead us to postulate *exactly* this state of the radioactivity evidence. We would expect radiogenic minerals to indicate very large ages and we would expect different elements in the same mineral, or different minerals in the same formation to agree with each other.[17]

Hundreds of years to lay the foundation for modern geology, the discovery of radioactivity, and the development of technology for radiometric dating, and these folks say that they knew what the answer would be all along? Something funny is going on here. Something that illustrates *exactly* what such people think of science and our ability to understand the natural world around us.

To appreciate what they are doing, we have to journey back 10,000 years or so to the morning of their supposed creation day. In their view, the universe was fully formed at the instant it was willed into being. Eden was a large and mature garden. Its landscaping did not look like the newest house in the subdivision, with a half-grown lawn and tiny saplings fresh from the divine nursery. Rather, its trees and shrubs and animals were fully grown from the instant of creation. They had, in the words of creationists, an "appearance of age."

The appearance of age, by their logic, must have applied to everything, including the geology of the planet and even the cosmos:

> Both parent and daughter elements in each radioactive chain were created at the beginning, probably in "equilibrium" amounts. The amount

> of originally created radiogenic end-product in each chain is uncertain; it is likely, however, that homologous amounts were created in all such minerals so that all such elements would, when created, give an "appearance" of the same degree of maturity or of age.[18]

The key statement in this passage is that all elements would "give an 'appearance' of the same degree of maturity or of age." In other words, if you're going to create radioisotopes and all of their potential decay products fully formed on creation morning, you have to decide what proportions of all of those decay products will be present in the minerals of your universe. And, lest you, the Creator, be thought of as confused and disorganized, all those minerals should all be set to the same radiometric clock. Whitcomb and Morris say this explicitly:

> It is more satisfying teleologically, and therefore more reasonable, to infer that all these primeval clocks, since they were "wound up" at the same time, were also set to "read" the same time.[19]

Now things get really dicey. Those lions and tigers and bears in the Garden of Eden might have just been adults of any age, but the grim precision of radioactive decay requires that a *specific* age was programmed into the materials of the planet—and so it was, according to the creationists. The Creator didn't just make things look old. He made things look as though they were a *specific* age. In the case of the earth and its solar system neighbors, this means that He intentionally fashioned their materials to look as if they were 4.5 billion years old when in fact they were brand-new.

With the great advances in astronomy of recent years, we routinely analyze stars and galaxies as far as 8 or 9 billion light-years from the earth. As astronomers emphasize, collecting the light from such great distances means looking into the past. When we image a galaxy several billion light-years distant, we are not looking at that galaxy as it is right now; we are looking at how it appeared billions of years ago.

How do the young-earth creationists handle this problem? Once again, they invoke the ingenuity of the Creator:

> It [the universe] must have had an "appearance of age" at the moment of creation. The photons of light energy were created at the same instant as the stars from which they were apparently derived, so that an

observer on earth would have been able to see the most distant stars within his vision at that instant of creation.[20]

Quite a picture, isn't it? Adam looking into the night sky on the first day of his existence, contemplating the beauty of thousands of stars, despite the fact that the nearest star other than the sun is more than four light-years away. The Creator, clearly, didn't want him to have to wait four years to enjoy that first nighttime star, so he made all of the intervening photons at once.

This may sound reasonable, but consider the implications. In the 10,000 years since creation, the *actual* starlight that has had enough time to reach us comes from only a tiny proportion of our neighbors. This means that every event witnessed at a distance by the Hubbell space telescope and other astronomical instruments, including the explosive disintegrations of stars and the gravitational effects of black holes, is fictitious. None of these things really happened—they were all constructed, artificially, in the trillions of photons assembled by the Creator to give His cosmos an appearance of age.

There is no way around this problem. One can reconcile a recent creation with the size of the known universe in only two ways: one, by fooling with the fundamental constants of nature; or two, by requiring that every astronomical object and event more distant than 10,000 light-years is fictitious. Some creationists have opted for the first alternative,[21] taking on the most fundamental constant in the universe, the speed of light. They claim that it was *much* faster in the past, and that accounts for the light from distant galaxies that is now reaching us from billions of light-years away. The lack of evidence does not seem to bother them, but it certainly will trouble physicists. As a number of my colleagues have written, an assault upon the scientific integrity of one field of science (biology) quickly becomes an assault upon all of science—in this case, taking down astronomy and physics to fashion a case against evolution.

To my way of thinking, the second alternative is the dangerous one. After all, an argument for the inconsistency of the speed of light can be refuted by the simple observation that its speed truly is constant, and that's that. The second alternative requires that the Creator of the universe intentionally fashioned a bogus astronomical history extending as far into space as our instruments can probe. And that's not all.

He also set those radiometric clocks to an apparent date for the creation of the solar system of 4.5 billion years, a 40 million percent exaggeration of its actual age, according to creationists. To me, this sounds like a deception most cruel. *Their* Creator deliberately rigged a universe with a consistent—but fictitious—age in order to fool its inhabitants.

Clearly the creationists recognize this problem, and have already tried to escape from this predicament through the trapdoor of faith:

> Whatever this "setting" was, we may call it the "apparent age" of the earth, but the "true age" of the earth can only be known by means of divine revelation.[22]

I do not dispute the fact that many people find what they believe to be divine revelation preferable to scientific knowledge. Our modern-day creationists are certainly not the first people in history to make that choice, although ironically they may be the first to invoke the name of science itself, as in "scientific creationism," even as they reject science. This lack of honesty is most revealing.

What saddens me is the view of the Creator that their intellectual contortions force them to hold. In order to defend God against the challenge they see from evolution, they have had to make Him into a schemer, a trickster, even a charlatan. Their version of God is one who intentionally plants misleading clues beneath our feet and in the heavens themselves. Their version of God is one who has filled the universe with so much bogus evidence that the tools of science can give us nothing more than a phony version of reality. In other words, their God has negated science by rigging the universe with fiction and deception. To embrace that God, we must reject science and worship deception itself.

On a scientific basis, the claims of the creationists are especially easy to refute. Most scientists, quite rightly, have ignored the religious claims of the creationists, but those claims are worth noting if only to emphasize the insidious danger they present to *both* science and religion. One can, of course, imagine a Creator who could have produced all of the illusions that the creationists claim to find in nature. In order to do so, we must simultaneously conclude that science can tell us nothing about nature, and that the Creator to whom many of us pray is inherently deceitful. Such so-called creation science, thoroughly analyzed, corrupts both science and religion, and it deserves a place in the intellectual wastebasket.

4

GOD THE MAGICIAN

For many critics of evolution, the intellectual price demanded by scientific creationism is just too high. This is not because of lingering sympathies for the ideas of Charles Darwin or those who have followed him. And it certainly is not due to a lack of ideological kinship among young-earthers like Henry Morris and Duane Gish and other anti-evolutionists who do not subscribe to the complete creationist vision. It is because young-earth creationism requires a full frontal assault on virtually every field of modern science.

Although biologists are the principal targets of this attack, the intellectual need to twist every scientific observation into a creationist framework means that no science is safe. As we have seen in the previous chapter, geological dating of rocks is ignored, astronomy is reduced of its fundamental constants, and even chemistry is subverted by a creationist requirement that the second law of thermodynamics be stretched to prevent the evolution of complex systems.[1]

For most reasonable people living in a technological society at the transition point between two great scientific centuries, the frank invalidation of so much science is just too great an outrage. Whatever the flaws and limitations of scientific enterprise, its impressive productivity makes an outright rejection of so much evidence simply nonsensical to most people. As a result, even the critics of evolution are increasingly unwilling to believe that the whole of science is organized into a grand conspiracy against them.

This, I suppose, is a kind of progress; but more to the point, it also makes good strategy. If they limit their attack on evolution to just a few subdisciplines of one field, they may convince a science-respecting public that most of science is just fine, and it's just one subfield of a single science, biology, that has strayed.

If they apply this strategy carefully, they might even claim to be the only true friends of authentic science. They might agree that the current scientific description of the world, its history and chemistry and biology, are *essentially* correct. In the name of objectivity, scientific rigor, and open-mindedness, they can then make the case that in one particular area, nearly a century and a half ago, some poor fellow named Darwin got the origin of species persuasively, dramatically, and overwhelmingly wrong.

The historical consequences of Darwin's great mistake would be undeniable, so the thinking would go. If they could demonstrate the flaws in his thinking and carefully extract them from current biological dogma, they just might be able to excise the tumor of Darwinism and leave an otherwise valuable body of biological science alive and intellectually well.

To pull off this coup de grace, they would first need an opening, a chink in the armor of evolution's consistent and powerful grip on so many lines of evidence. They would have to find something that doesn't seem to fit. A neophyte critic of science might think that this could best be done from the outside, by developing independent lines of research, criticism, and interpretation. Those who understand the scientific process know better. They appreciate the fact that the best way to find allies to attack *any* idea is to look within the scientific establishment itself, since the nature of science all but guarantees that such opportunities will abound.

Did Darwin Get It Wrong?

If scientific critics of Darwin were difficult to find in the sixties and seventies, by the early eighties they couldn't be missed. The headline of one particularly memorable new article in the journal *Science* proclaimed, "Evolutionary Theory Under Fire."[2] It took a few years for the ferment to make its way through the scientific food chain into the popular press.

By 1986, no less an intellectual figure than neoconservative editor and critic Irving Kristol would proclaim:

> Though this theory [the neo-Darwinian synthesis] is usually taught as an established scientific fact, it is nothing of the sort. . . . In addition, many younger biologists (the so-called "cladists") are persuaded that the differences among species—including those that seem to be closely related—are such as to make the very concept of evolution questionable.[3]

How did the "very concept of evolution" come to be called questionable? The controversy began innocently enough, as many scientific ideas do—in the heads of a couple of young scientists, Niles Eldredge and Stephen Jay Gould, whose energetic and accomplished careers teach an important lesson about the scientific profession. "Don't rock the boat" might be good advice for an organization man trying to climb the corporate ladder, but for a scientific career, that slogan does not work. In science, where confirmatory data and consistency with established theory are regarded as boring, the best advice would be to rock the boat vigorously—maybe even to turn the damn thing over. That's exactly what Eldredge and Gould managed to do.

In a 1972 paper, tucked away in a book on the evolutionary subfield of paleobiology, they took issue with what they considered the prevailing view of the pattern of evolutionary change over time. That view was the "modern synthesis" of evolutionary theory. English biologist Julian Huxley had coined the term in 1942 to describe the powerful and productive synergies that had developed between evolution, the new science of genetics, and the emerging field of population biology. Eldredge and Gould read the modern synthesis to mean, among other things, that evolutionary transformations had to be gradual, and that natural selection steadily pulled the characteristics of a population in one direction or the next, producing a record of continual transformation over time.

Darwin himself had endorsed the view that evolutionary change must be gradual, and underlined this viewpoint in a famous sentence in *The Origin:*

> If it could be demonstrated that any complex organ existed, which could not possibly have been formed by numerous, successive, slight modifications, my theory would absolutely break down.[4]

Whatever one's view of the original Darwin, I agree that a fair reading of "numerous, successive, and slight" allows "gradual" as an appropriate interpretation. Therefore, any novel organ or structure appearing in evolutionary history ought to be explicable in terms of natural selection working gradually on preexisting structures. A bit later, Darwin reiterated this point by quoting the maxim, *Natura non facit saltum* (Nature doesn't like to jump).[5] And did his best to make himself perfectly clear once again:

> Why should not Nature take a sudden leap from structure to structure? On the theory of natural selection, we can clearly understand why she should not; for natural selection acts only by taking advantage of slight successive variations; she can never take a great and sudden leap, but must advance by short and sure, though slow steps.[6]

The modern synthesis extrapolated Darwin's clarity on this point to endorse a pattern of evolutionary change in which over time, natural selection alters the average characteristics of a population in one direction or another, eventually giving rise to a new and distinct species. Critics of this theory called it "phyletic gradualism." Eldredge and Gould said that phyletic gradualism would require that a graded series of intermediates exist between the two species, and that only by finding many such series in the fossil record would modern synthesis be proved correct.

They proclaimed that paleontologists had known for years that this was *not* what the fossil record looked like. The popularizers of the modern synthesis had been glossing over the principal, overwhelming characteristic of the fossil record—*stasis*. As Gould said:

> For millions of years species remain unchanged in the fossil record, and then they abruptly disappear, to be replaced by something that is substantially different, but clearly related.[7]

Eldredge and Gould coined their new idea "punctuated equilibrium," and presented it as a radical alternative to the more traditional gradualism that had been taken as the evolutionary norm.

By 1980, punctuated equilibrium had become *the* hot topic in evolutionary theory, and by 1990 it had risen to the status of textbook orthodoxy. When Joseph Levine and I wrote our first college textbook on general biology, we felt compelled to include extensive coverage of punctuated equilibrium, to seem up to speed with the fashions of movers and shakers in evolutionary biology.

Punctuated equilibrium's advance to acceptance was so spectacular that Gould and Eldredge were able to celebrate the twenty-first anniversary of their brainchild in a retrospective article in *Nature* in 1993. They characterized it as

> a novel interpretation for the oldest and most robust of paleontological observations: the geologically instantaneous origination and subsequent stability (often for millions of years) of paleontological "morphospecies."[8]

Instantaneous origination, followed by stasis. Living things persist for millions of years in an *equilibrium* of ecological relationships and communities. Then, from time to time, that static equilibrium is *punctuated* by brief periods of rapid change. In graphic terms (see Figure 4.1), evolution via punctuated equilibrium is represented as a branching bush, in which the vertical lineages change little over time, giving rise to new, equally static branches produced in a geological instant.

In verbal shorthand, the concept of gradual change over time in a single lineage became known as "Darwinian" change. By contrast, the sudden appearance of a new species represented "punctuation." Finding one fossil sequence after another that seemed to match their conception of change and stasis, Gould and Eldredge proclaimed much of the fossil record to be "non-Darwinian." The implication was that Darwin got it wrong, and they were able to quote chapter and verse from Darwin himself on the gradual nature of change required by *his* theory of evolution. The emerging data on punctuation lent support to the idea that "punk eek" was something genuinely new.

The prominence of punctuated equilibrium in scientific and popular thinking was heightened immeasurably by the literary works of Stephen Jay Gould. His monthly essays on evolution, which appear in *Natural History* magazine, are surely among the finest examples of science writing of this or any other time. His ability to weave the currents of many cultures, from French poetry to popular cinema to baseball, into his writing is matchless. A passion for life and for the richness and complexity of the fabric of evolution is everywhere in his work. The sheer

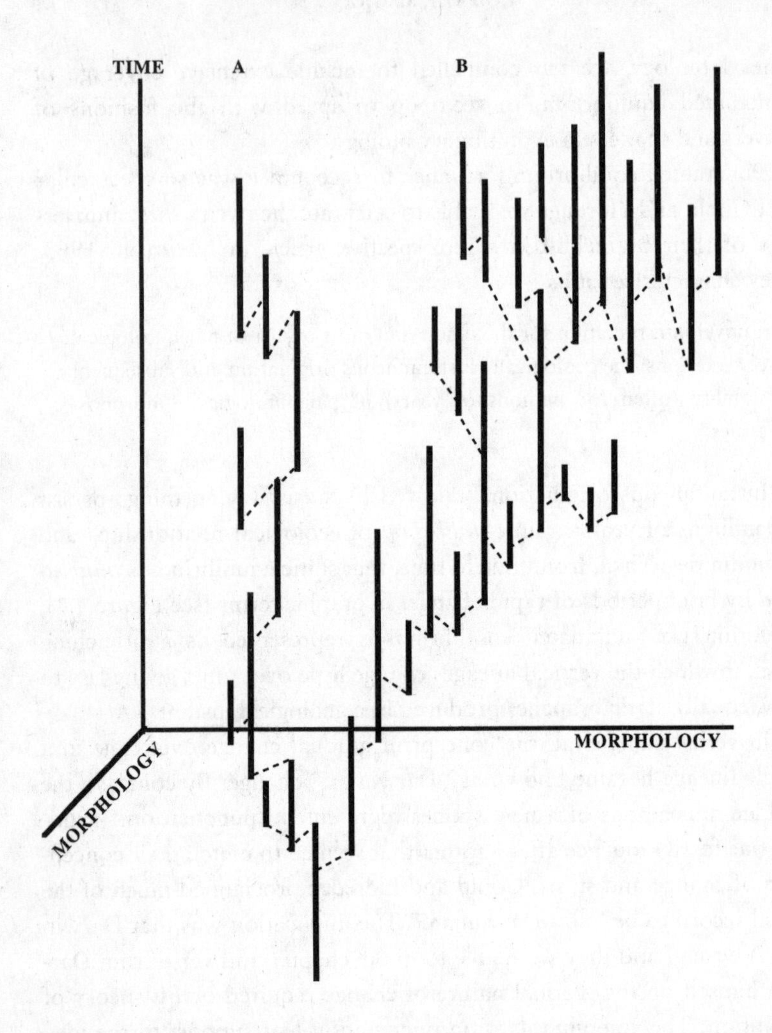

Figure 4.1. Punctuated equilibrium as a model of evolutionary change over time. This diagram shows two model lineages. Lineage A shows long-term stasis, while lineage B shows a persistent pattern of speciation and evolutionary innovation. Individual species are depicted as vertical bars that persist briefly in the fossil record, punctuated by rapid transitions (dotted lines) that result in new species.

Redrawn from S. J. Gould and N. Eldredge, "Punctuated Equilibrium Comes of Age," *Nature* 366 (1993): 223–227.

strength of his prose has helped to ensure that punctuated equilibrium would occupy an important place in the scientific consciousness.

Despite Gould's efforts, however, punctuated equilibrium remains a controversial, even contentious, topic among evolutionary biologists. Two diagrams (see Figure 4.2), redrawn from science writer Roger Lewin's important summary of a 1980 meeting on evolutionary biology, hint at some of the reasons for this. We will return to those reasons later in this chapter.

Whatever the merits or imperfections of this view of evolutionary change, as a generalization of the nature of the fossil record, it pointed to two ideas that had not been part of the common thinking about natural history prior to Gould and Eldredge. The first was that the fossil record itself was non-Darwinian, that it somehow did not match Darwin's expectations for the evidence required to validate his theory. The second was that new species might originate in an instantaneous burst of rapid, horizontal genetic change, rather than from the "normal" interplay of variation and natural selection that dominates long periods of stasis.

Gould and Eldredge may have seen these ideas as new ways to enhance our understanding of the mode and tempo of evolutionary change, but the opponents of evolution saw them as something else—the long-sought opening through which the forces of reason could attack the Darwinian dogma.

Reasonable Doubt

As Gould's marvelous books and essays streamed forth, the Harvard professor quickly became the single most recognizable voice for evolution in the United States, and quite possibly in the world. When an expert witness was needed for a public trial in which the scientific status of evolution had been called into question, Gould was there to testify eloquently and effectively. If a national news program needed someone to speak with precision on the meaning of a new fossil discovery, or even of the possibilities for life on Mars, Gould was *the* expert. And when television producers needed a genuine scientist to narrate a natural history documentary with wit and genial authority, who better than Steve Gould?

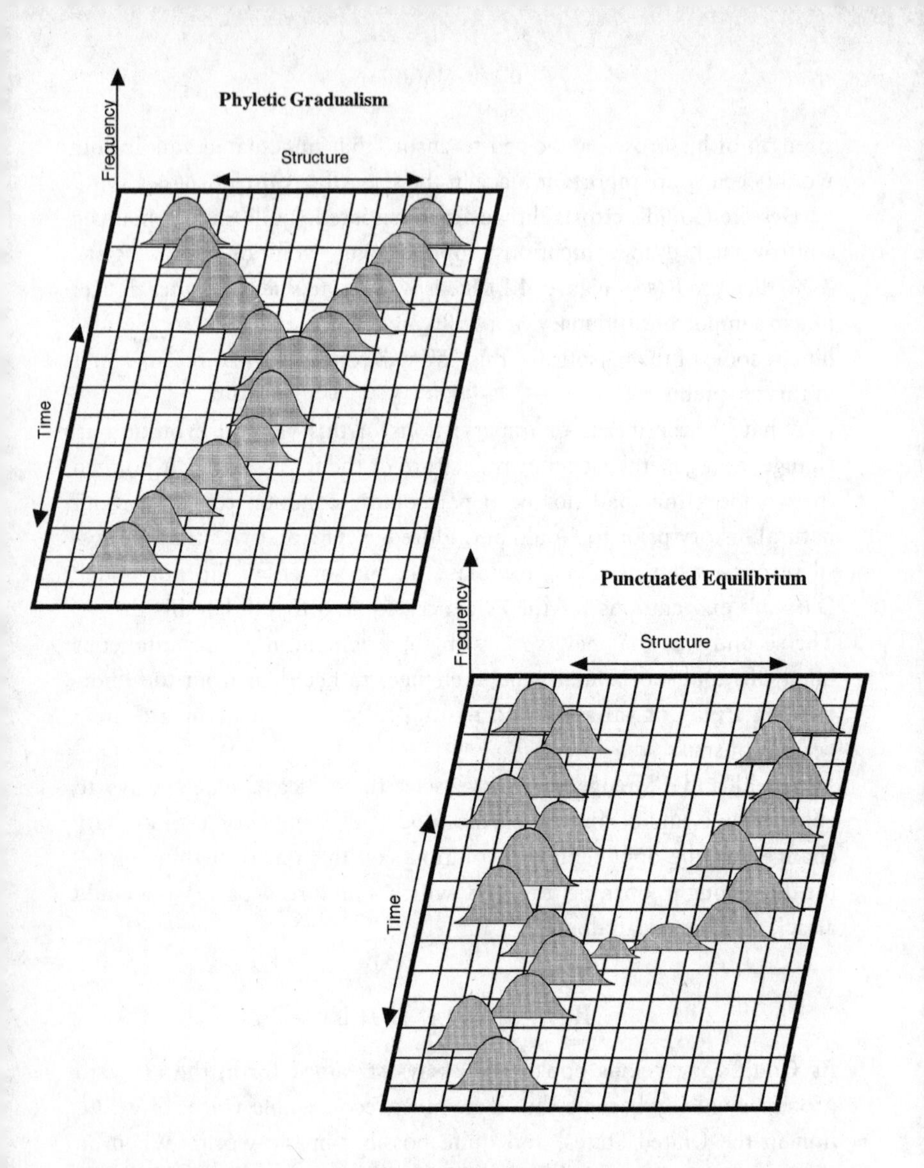

Figure 4.2. Phyletic gradualism (top) implies gradual change, but punctuated equilibrium (bottom) shows a more abrupt speciation process. The members of a population at each time point in the graph are shown as a Gaussian distribution representing the range of structural variation in key features. According to Gould and Eldredge, the divergence of one lineage into two is generally an abrupt event that "punctuates" long periods of stasis, or "equilibrium," during which little change takes place.

Diagrams redrawn from R. Lewin, "Evolutionary Theory Under Fire," *Science* 210 (1980): 883–885.

But prominence exacts a price, and one of the subtle costs of being transformed from revolutionary into establishment figure is that every comment is taken as representative of the core meaning of an entire field. If one wants to show what evolution "really means," a quotation from Stephen Jay Gould, now holding the status of official evolutionist number one, or Niles Eldredge, his coauthor of the revealed wisdom of punctuated equilibrium, will more than suffice.

The prominence of Gould and Eldredge in advancing punctutated equilibrium allowed the opponents of evolution to draw three conclusions:

1. The proponents of puncuated equilibrium, including Gould, Eldredge, and many others who speak for evolution, have said with authority that the fossil record is clearly non-Darwinian in character. Therefore, what had been taken as the centerpiece of evidence for the reality of evolution has now been shown, by mainstream scientists, to be nothing of the sort.

2. Only two things really *do* occur in the fossil record, according to punctuated equilibrium. Those are the instantaneous origination and subsequent stability of new species without any trace of ancestors.

3. Although scientists have demonstrated genetic mechanisms that can produce gradual changes within species ("microevolution"), no known genetic mechanism can produce the radical changes ("macroevolution") required for the sudden appearance of new species.

To critics of evolution, words like "sudden appearance" must have had a marvelous ring. Wasn't that *exactly* what should be expected for the special creation of a species? If new species, once they were specially created, showed no change over time, wouldn't that confirm that the kind of evolution our kids are taught in schools never actually happened?

One might conclude that when evolutionists claim to be able to explain the origin of species as a result of natural, material processes observable in the field and in the laboratory, they are engaged in a tactic of deceit, covering up the lack of any real evidence for evolution. Rushing to exploit exactly this opening, Berkeley law professor Phillip Johnson gleefully quoted one of the founders of punctuationalism:

> Niles Eldredge has written: "No wonder paleontologists shied away from evolution for so long. It never seems to happen." New things appear suddenly in rocks dated in different ages, but there is no pattern of gradual transformation and no ability to identify specific ancestors of major groups. Although Eldredge admits that the fossil record contradicts the theory of gradual adaptive change, he nonetheless calls himself a "knee-jerk neo-Darwinist," meaning apparently that he believes the theory despite what he knows as a paleontologist.
>
> The non-occurrence of Darwinian change is particularly evident where fossils are most plentiful—in marine invertebrates, for example. There it's all variation within the type, with no substantial evolution.[9]

Putting Johnson's argument another way, the evolution that biologists say they can demonstrate as scientific fact does not fit the sudden punctuation that the fossil record demands. So maybe, just maybe, Darwinian evolution is flat-out wrong!

Johnson's assault on Darwinism carries the unmistakable marks of his profession. In a 1991 book, *Darwin on Trial,* Johnson asked the question:

> Can something be non-science, but true, or does non-science mean non-*sense*? Given the emphatic endorsement of naturalistic evolution by the scientific community, can outsiders even contemplate the possibility that this officially established doctrine might be false? Well, come along and let us see.[10]

Although he is a non-scientist, Johnson's academic status as tenured law professor at a major university gave his questions about evolution a respectability that other critics might well have envied.[11] He skillfully attacked evolution as an "establishment" theory, expertly playing the role of outside crusader for truth and justice who marshals the best instincts of the American legal profession.

Johnson exploits the punctuational critique of Darwinian change to attack evolution itself. The real fossil record shows jumps and sudden appearances, and since no genetic mechanism can explain such jumps, goes the argument, the whole idea of evolution is wrong.

The specifics of Johnson's *scientific* criticism of evolution aren't novel at all. (His critique includes the additional *philosophical* notion that evolution is the core of an anti-spiritual materialism. This idea is the key to understanding the motivation behind Johnson's crusade, and will be examined in Chapter 6, "The Gods of Disbelief.")

First, he argues that the ancestor-descendent relationships apparent in the fossil record do not exist—the gaps between individual species in such supposed ancestral sequences are just too great. The standard that Johnson applies in his analysis is that a direct, provable ancestral line of descent must be established. Without such a line, fossil evidence, however detailed, is irrelevant to the validity of evolution.

To cite just one example, he argues against the growing documentation on the evolution of land vertebrates, or tetrapods, from fishes by quoting from Barbara J. Stahl's textbook on vertebrate history that "none of the known fishes is thought to be directly ancestral to the earliest land vertebrates."[12] In so doing, he overlooks compelling evidence, also described by Stahl, that the first tetrapods were, in fact, descended from fish in favor of the much smaller point that the *exact* fossil fish ancestral to the first tetrapods may not yet have been discovered. Similar dismissals are crafted against several other lines of descent that are generally taken as strong evidence for evolution.

Second, he contends that evolution has a problem of mechanism. Here he plays directly upon Gould and Eldredge, demanding a genetic mechanism that could accomplish the jumps in form and function that punctuated equilibrium demands. Jumps that, Johnson believes, are impossible.

The gut reaction of just about any scientist familiar with the evidence supporting evolution would be to answer Johnson's two criticisms head-on. No credible evolutionary sequences? "Those fossil sequences do support evolution, and here are five or ten or twenty examples!" would be the reply. No mechanism? "You've got to be kidding? We can measure the actual rate of morphological change produced in the real world by natural selection, and guess what? It turns out to be ten to a hundred times *faster* than the amount required to explain even the most rapid transitions in the fossil record!"[13] Johnson's criticisms of historical evolutionary sequences do not hold up under scrutiny, and his charge of a missing mechanism is absolute nonsense, as we will see.

But like any good defense lawyer, Johnson doesn't believe that he has to prove *his* case. All he needs is reasonable doubt. If he can punch a few holes here and there in evolutionary theory, he can hope that the jury of public opinion will hold evolution to be something less than fact. And that—make no mistake about it—is his real goal.

Johnson and his colleagues deserve to be taken seriously, and that is exactly what I propose we do—more seriously, in fact, than they take themselves. Let us accept their proposals as serious suggestions for the natural history of planet earth, and see where they lead. This is something that Johnson never does—and there is a very good reason why.

The Sorcerer's Apprentice

A little knowledge is a dangerous thing.

Like the sorcerer's apprentice watching the master magician, Johnson has watched evolutionary biologists at work. As he has noticed, they use Darwin's natural selection to account for the origin of species. Objecting to natural selection's materialist character, he gets an idea. He'll substitute "intelligent design" for natural selection, and *voila*—the fabric of natural history remains the same, but now it can be linked to an unnamed master designer.

As his fellow evolution critic David Berlinski puts it:

> The structures of life are complex, and complex structures get made in this, the purely human world, only by a process of deliberate design. An act of intelligence is required to bring even a thimble into being; why should the artifacts of life be different?[14]

By such logic, Johnson and Berlinski attribute the intricate structures and stunning diversity of life to intelligent design. Not unlike the apprentice who believed he understood his master's magic, they may have thought this a simple alternative to evolution; but they might have taken a lesson from the fate of the apprentice in that story. Before long, his novice efforts spun out of control, and the very spells that he had cast in triumph turned against him with a vengeance. Intelligent design is such a spell.

When the critics of evolution invoke design to replace evolution, no doubt they believe they have solved their problems at a stroke. A divine designer does not need a mechanism. Such a designer can, by definition, create *anything*. There is no fossil, no intermediate form, no evidence of any sort that cannot be explained away by invoking intentional, intelligent design.

If we were to stop right there, things just might work out. As a purely rhetorical alternative to evolution, intelligent design serves the defense team well. It raises the specter of reasonable doubt, and allows its advocates to claim that those on the other side of the table are closed-minded. After all, unlike young-earth creationists, intelligent-design theorists do accept the scientific work of physics and geology on the age of the earth, and they further agree that the patterns of the fossil record are a genuine historical sampling of the nature of life in the earth's long past. Indeed, they might claim to have followed the findings of every scientific field but one, dissenting from evolution only because they are not bound by the constraints of materialism. This, of course, allows them to be open-minded enough to consider the possibility of design. And what's the matter with that?

The matter is that, if Johnson is right, then we should apply the explanation of design to every event in the natural history of the planet. It's not logically tenable to allow that evolution could have produced some species but not others; therefore, the explanation of design must be invoked for the origin of every species. The modern advocates of this idea seem to have forgotten that at one time that is exactly what biologists did. Until, of course, it became impossible to take seriously.

The trouble begins with little things. Consider two isolated groups of islands, for example: Cape Verde, off the coast of Africa, and the Galapagos, off the western coast of South America. Both of these island groups contain species that are endemic to them; that is, they are found nowhere else in the world. Such species, like all organisms, must have been intelligently designed, according to Berlinski and Johnson. One might wonder *why* a designer would go to the trouble of crafting several dozen unique species just for each of these tiny archipelagos, but let's set that question aside for a more telling one. Why was the designer's powerful imagination so limited by the islands' neighborhoods? As Darwin himself noted:

> There is a considerable degree of resemblance in the volcanic nature of the soil, in climate, height, and size of the islands, between the Galapagos and Cape de Verde Archipelagos: but what an entire and absolute difference in their inhabitants! The inhabitants of the Cape de Verde Islands are related to those of Africa, like those of the Galapagos to America. I believe this grand fact can receive no sort of explanation on the ordinary view of independent creation.[15]

In short, the unique species of each island group are closely related to species on their respective, somewhat distant mainlands. The question begged by this situation is *why*.

Why would species that had been designed just for the Galapagos archipelago, and nowhere else on earth, bear so plainly the stamp of affinity to those designed for South America? The advocates of intelligent design have no explanation beyond the whim of the designer himself. That's just the way he chose to do it.

Evolution offers the perfect explanation. Both groups of islands are geologically recent. A few founding species from the respective mainlands colonized each archipelago, and then geographic isolation allowed natural selection to go to work. Therefore, the new species that developed on each island group are the descendants, modified by evolution, of the respective founding populations from each continent. That's why each archipelago has its own collection of unique species, and why those species most closely resemble their respective nearby mainlands.

In following the intelligent design argument, one would have to believe that it was also the designer's choice to mislead—by producing sequences of organisms that mimic evolution so precisely that generations of biologists would be sure to misinterpret them. Consider, for example, an animal that God Himself (as quoted in the Book of Job) considered one of His "masterpieces":

> *His tail is like a cedar.*
> *The sinews of his thighs are knit together.*
> *His bones are like brass cylinders;*
> *His ribs are like bars of iron.*
> *He is the masterpiece of My animal creation:*
> *The One that made him furnished him alone with a sword.*
>
> JOB 40: 17–19

The "animal" in which God took such pride was, of course, the elephant. To someone not familiar with the fossil record of the elephant—which is to say just about everyone—the claim that *Loxodonta africana*, the African elephant, and *Elephas maximus*, the Indian elephant, were intelligently designed might seem attractive. An intelligent designer could indeed produce a skull that counterbalanced the mass of the trunk, fashion the "sword" described by Job, and do the structural engineering required to thicken the bones of the elephant's leg to support its great weight.

So elephants, like every other intelligently designed animal, must appear suddenly in the fossil record, fully formed and without ancestors. Right?

Well, not exactly. Courtesy of the fossil record, the apprentice's half-baked magic is about to spin out of control.

The skulls, teeth, and jaws of elephants are distinctly different from other mammals, which make extinct elephant-like organisms easily recognizable from fossils. In 1997, Hezy Shoshani, the founder of the Elephant Research Foundation, described some of these extinct proboscideans (elephant-like animals) in *Natural History*.[16] Beginning in the Eocene, more than 50 million years ago, he traced the evolution of the two distinct species of modern elephants. Summarizing this data for the reader (see Figure 4.3), Shoshani presented the kind of branching lineage that should be familiar to anyone who has looked into the geological record of any living species.

Like it or not, intelligent design must face these data by arguing that each and every one of these species was designed from scratch. For some reason, then, that great designer first engineered a small trunk into a little critter called *Paleomastodon* at the beginning of the Oligocene some 35 million years ago. Ten million years later, the trunk design was used again in the larger *Gomphotherium,* along with a set of protruding tusks. Evidently the designer now thought that the trunk was a good idea, because he used it again in *Deinotherium* and *Platybelodon* in North America, and for *Gomphotherium* in Africa, all at the beginning of the Miocene. By the end of the Miocene, *Primelephas,* whose well-developed trunk and tusks are unmistakably similar to the larger species of modern elephants, would also appear in Africa.

The hypothesis of design absolutely, positively requires the successive creation of increasingly elephant-like organisms over time. If some of the new "inventions" of structure and skeleton in *Paleomastodon* seem to be carried over in *Gomphotherium,* don't be misled. It's not because of ancestry. One organism has nothing to do with the other. Their similarities and their differences are coincidental, just like their successive appearances over geologic time.

Now, I can imagine Phillip Johnson pointing to Figure 4.3 and telling

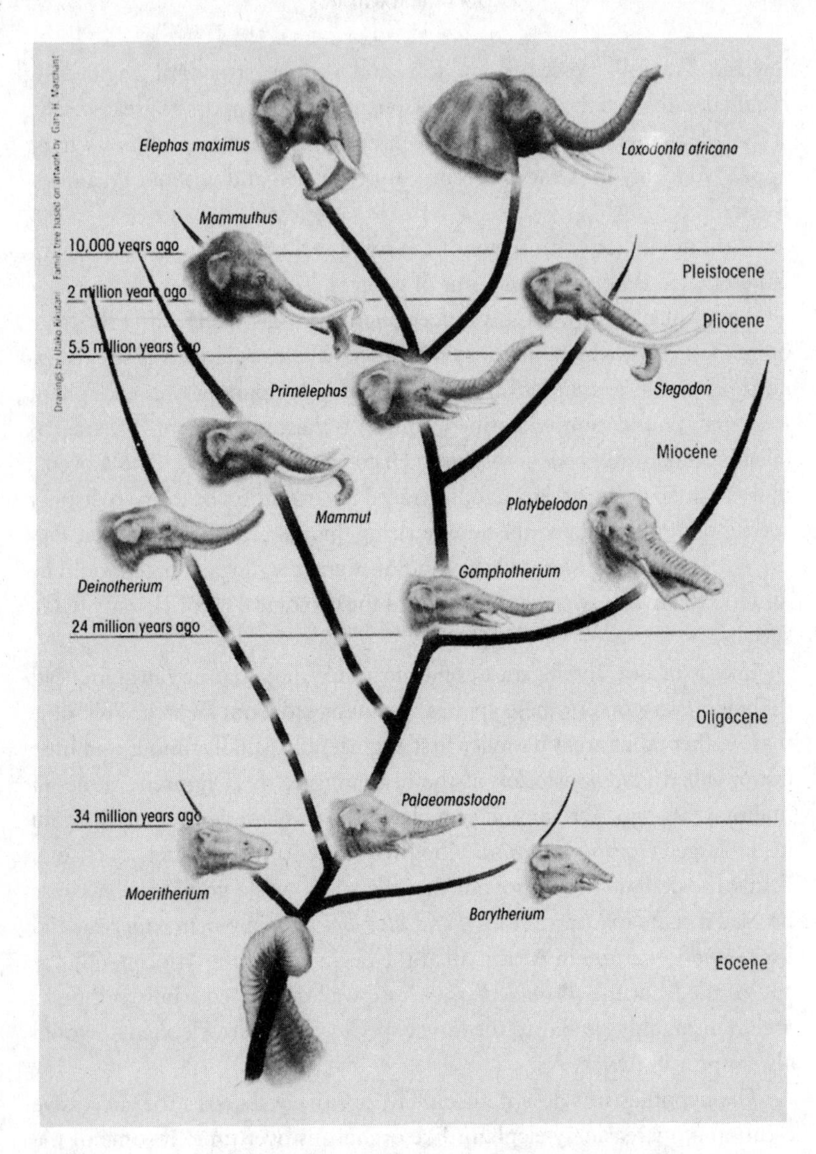

Figure 4.3. A simplified family tree showing representative members of the elephant, or proboscidean, lineage. Like most modern mammals, the two extant species of elephants are recent appearances, both having arisen within the last 2 million years. The fossil record of their ancestors, now increasingly detailed, goes back nearly 50 million years, well into the Eocene.

Diagram from J. Shoshani (1997) *Natural History*, 106 (10), November 1997, p. 38.

me, with a straight face, that the top part of Shoshani's evolutionary tree might indeed be the product of design. First the designer experimented with one species (*Primelephas*), and then he replaced it with three. Of those three, one (*Mammuthus*, the woolly mammoth) became extinct, and two survive to the present day. Each was a perfectly designed and perfectly formed species, and the sequence of their appearances is a misleading coincidence. Just like the further coincidences that show ancestry of the elephants way back into the Eocene.

Unfortunately, the fine details of fossil reality are even worse for the argument of design. As with most popular representations of natural history, Shoshani's family tree is a gross oversimplification. Those four organisms near the tree's apex are actually four genera (see Figure 4.4a), meaning four groups of closely related species. When one plots the species-specific fossil record of the elephants (Figure 4.4b), it reveals not four, but twenty-two distinct species in just the last 6 million years.

This designer has been busy! And what a stickler for repetitive work! Although no fossil of the Indian elephant has been found that is older than 1 million years, in just the last 4 million years no fewer than nine members of its genus, *Elephas,* have come and gone. We are asked to believe that each one of these species bears no relation to the next, except in the mind of that unnamed designer whose motivation and imagination are beyond our ability to fathom. Nonetheless, the first time he designed an organism sufficiently similar to the Indian elephant to be placed in the same genus was just 4 million years ago—*Elephas ekorensis*. Then, in rapid succession, he designed ten (count 'em!) different *Elephas* species, giving up work only when he had completed *Elephas maximus,* the sole surviving species.

A reasonable person, eager to accept intelligent design as an explanation for this one living species, must therefore believe that the designer started more than 50 million years ago with a small organism quite unlike the Indian elephant. Then, over time, he crafted scores of new species, his designs gradually drifting closer and closer to the modern elephant. In the last few million years, he constructed in rapid succession nearly a dozen semifinal drafts until *Elephas maximus* finally came off the drawing table.

This curious pattern of design that resembles succession is repeated countless times in the fossil record; and for each instance, Johnson,

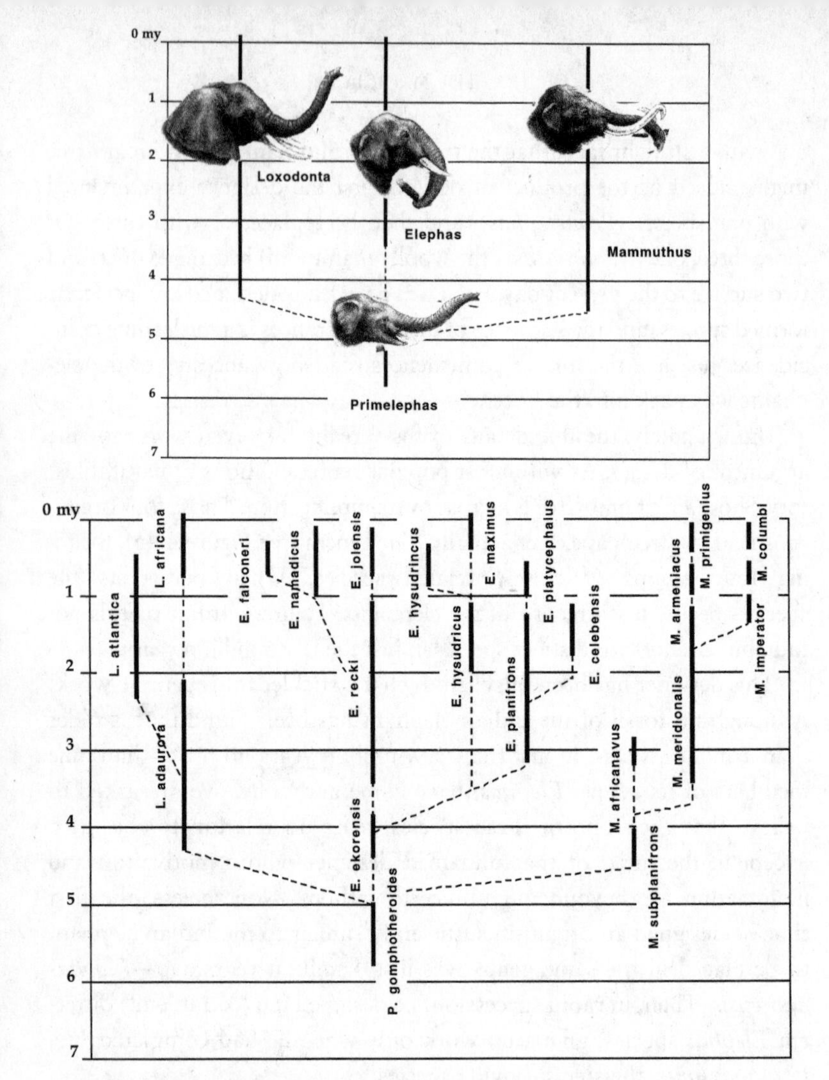

Figure 4.4a and b. A close analysis of the last 6 million years of elephant evolution shows the depth and complexity of the fossil record. Two modern and one recently extinct form (the woolly mammoth, *Mammuthus*) all trace their ancestry to *Primelephas*. The comparatively simple pattern for these three genera (top) belies the fact that twenty-two new species have appeared within them over the past 4.5 million years.[17] The classic branching of this lineage is a typical evolutionary pattern. Intelligent design, however, regards the appearance of each new species as the result of an individual act on the part of a designer.

This diagram has been redrawn from Figure 15 of V. J. Maglio's "Origin and Evolution of the Elephantidae," *Transactions of the American Philosophical Society* 63, no. 3 (1973): 1–149.

Berlinski, and their colleagues must claim that any impression of a sequence is just a figment of our imagination.

Is it any wonder that biologists are unable to take intelligent design seriously?

Over and over again, the imposition of intelligent design on the facts of natural history requires us to imagine a designer who creates successive forms that mimic evolution. Magicians are master illusionists, and if this magical designer had anything in mind, it must have been to cast the illusion of evolution and nothing else.

A Puff of Smoke

Intelligent design was the first, the original, and the simplest answer for the exceptional adaptations of living things to their environment. It was only when Charles Darwin came upon the scene that this simple explanation was abandoned. As Johnson puts it:

> The answer Darwin came up with was that these adaptations, which had seemed to be intelligently designed, are actually products of a mindless process called natural selection.[18]

Johnson, of course, favors the original explanation. If he is right, we should easily be able to fit the explanation of design into every aspect of the fabric of natural history. Evolutionary biologists do this all the time, seeking correlations between evolutionary trends and global changes in climate, for example, or with local events such as floods and food shortages. But I have never read, nor do I ever expect to read, an exploration of any event in natural history in which the explanation of design is correlated with actual events. It's easy to understand why.

New species appear constantly throughout the fossil record. If they really are the products of intelligent design, then how would such new species appear? I can think of only two possibilities, which are two more than the proponents of this theory are willing to present. First, the new organism appears magically, coming into existence in a literal puff of smoke. Second, the new organism is born as the *apparent* offspring of another species, but is so distinctive genetically that it becomes the founding member of a completely new species.

Don't think for a second that either of these would be rare occurrences. Twenty-two new species have arisen in the elephant lineage in the last 5 million years, roughly one every 230,000 years. And that's just for elephants. There are approximately 10,000 living species of mammals. If the rate of new elephant species is an indication, we might expect one new mammalian species to have popped up every twenty-three years or so. As rapid as this seems, these thousands of mammals are dwarfed by literally millions of insect species. To produce all those insects, the designer must have been even busier—that puff of smoke might have been a weekly occurrence.

Like it or not, intelligent design requires us to believe that the past was a time of magic in which species appeared out of nothing. That magic began with the dawn of life on this planet, and continued unabated for more than a billion years, bringing a grand parade of living things into existence. Throughout this time, novel organisms sprang into existence one after another, transforming the earth and producing eras in which organisms now long extinct dominated the planet.

So, which was it? Did the new species come into existence in a puff of smoke, or were they the genetically distinct offspring of another species? And, more important, why have they stopped? The appearance of new species out of thin air doesn't seem to happen anymore, even if it happened on a recurring basis in the geologic past. These newly designed organisms continued to appear on a regular basis right up to the present day. Then, for some reason, just as we became able to observe it, this remarkable magic stopped. Makes you wonder why, doesn't it?

The Question of Competence

Almost by definition, an intelligent designer would have to be a pretty sharp fellow. The critics of evolution like to emphasize the notion that the information found in biological systems must have been put there by an intelligent force. Hence, a designer, and a clever one, at that.

Biologists can have great fun with that notion. Living organisms, ourselves included, are loaded with what Stephen Jay Gould once called "the senseless signs of history." Our bodies do not display intelligent design so much as they reveal the evidence of evolutionary ancestry. Human embryos, for example, form a yolk sac during the early stages of development. In birds and reptiles, the very same sac surrounds a nutrient-rich yolk, from which it draws nourishment to support the growth of the embryo. Human egg cells have no comparable stores of yolk. Being placental mammals, we draw nourishment from the bodies of our mothers, but we form a yolk sac anyway, a completely empty one. That curiously empty yolk sac is just one sign that the ancestry of mammals could be found in egg-laying, reptile-like animals, a notion that is brilliantly confirmed by the fossil record.

These signs of history are the telltale marks of evolution, and all organisms have them. Because evolution can work only on the organisms, structures, and genes that already exist, it seldom finds the perfect solution for any problem. Instead, evolution tinkers, improvises, and cobbles together new organs out of old parts. A true designer would face no such problems, and could produce genuinely new structures, molecules, and organs whenever needed. Unfortunately for us, it doesn't seem to work that way.

The many imperfections of the human backbone which, regrettably, become increasingly apparent as we age, can hardly be attributed to intelligent design. They are easy to understand if we appreciate the fact that our upright posture is a recent evolutionary development. Evolution has taken a spinal column well adapted for horizontal, four-footed locomotion and pressed it into vertical, bipedal service. It works pretty well, but every now and then the stresses and strains of this new orientation are too much for the old structure. Intelligent design could have produced a trouble-free support for upright posture, but evolution was constrained by a structure that was already there. Chiropractors, of course, continue to reap the benefits.

To adopt the explanation of design, we are forced to attribute a host of flaws and imperfections to the designer. Our appendix, for example, seems to serve only to make us sick; our feet are poorly constructed to take the full force of walking and running; and even our eyes are prone to optical errors and lose their ability for close focus as we age. Speaking of eyes, we would have to wonder why an intelligent designer placed the neural wiring of the retina on the side facing incoming light. This arrangement scatters the light, making our vision less detailed than it might be, and even produces a blind spot at the point that the wiring is pulled through the light-sensitive retina to produce the optic nerve that carries visual messages to the brain.

We would also have to attribute every plague, pestilence, and parasite to the intentional actions of our master designer. Not exactly a legacy calculated to inspire love and reverence. Job's God may have bragged about the elephant, but would he really want to take credit for the mosquito?

Finally, whatever one's views of such a designer's motivation, there is one conclusion that drops cleanly out of the data. He was incompetent.

Careful studies of the mammalian fossil record show that the average length of time a species survives after its first appearance is around 2 million years. Two million years of existence, and then extinction. The story is similar for insects (average species duration: 3.6 million years) and for marine invertebrates (average duration: 3.4 million years).[19] In simple terms, this designer just can't get it right the first time. Nothing he designs is able to make it over the long term.

Not even those famous so-called living fossils, present-day organisms that are often said to have survived unchanged for millions of years, are exceptions to this rule. Horseshoe crabs, to pick one example, are indeed representatives of an ancient group, a superfamily known as the Limulacea, which extends back 230 million years to the early Triassic. Today's species, *Limulus polyphemus,* is a truly modern one that is easily distinguished from its Jurassic cousin, *Limulus walchi.*[20]

What is one to make of this? Quite simply, that the advocates of design are faced with a logical contradiction. They would like to claim that the perfection of design seen in living organisms cannot possibly have been achieved by a random, undirected process like evolution, and that an intelligent agent is required to account for such perfection. But when one looks at the record, the products of this intelligent design consistently fail to survive.

The actions this designer would have to take over time remind me of a story. As a visitor walked past one of the cages at a zoo, he marveled at the sight of a lion and a lamb sleeping peacefully next to each other. Amazed, he sought out the zookeeper. "That's incredible!" the visitor said. "How do you make them get along so well?" The zookeeper smiled. "It's easy. All we have to do is to put in a new lamb every day."

I wouldn't give that zookeeper an award for species conservation any more than I would give our hypothetical intelligence an award for design perfection. Not for a world in which ninety-nine percent of his creations have become extinct, not for a zoological park in which both lambs and lions must be restocked with completely new species time and time again, and certainly not for designing new life forms that consistently look like jerry-rigged modifications of his last creations. You might think the guy had no sense of originality. Worse yet, you might even think that evolution was going on.

THE "MISSING" MECHANISM

To many biologists, the most curious charge made by the intelligent-design camp is that the mechanisms thought to drive evolution are not up to the job. As Johnson puts it:

> Science knows of no natural mechanism capable of accomplishing the enormous changes in form and function required to complete the Darwinist scenario.[21]

That's a sweeping charge. If it were true, I and scores of other experimental scientists would walk over to Johnson's side of this issue in a flash. If modern work in molecular and developmental biology had indeed been unable to find any mechanism that could accomplish what Darwin required, it would be time to start thinking about an alternative. On the other hand, if modern researchers had discovered exactly such mechanisms, might it not be high time for critics to concede defeat?

One of the great discoveries of this century, courtesy of Watson, Crick, Franklin, and many others, is that genes are digital, which means that they are composed of a sequence of discrete units. Genes, the individual units of heredity, are written in the language of DNA, using a four-letter alphabet—A, C, G, and T—corresponding to the four nucleotides found in DNA: adenosine, cytosine, guanine, and thymine. The bacterium *Escherichia coli* contains 4,639,221 of these letters, organized into a total of 4,288 genes.[22] In groups of three, these letters code for other building blocks known as amino acids, which are strung together to make proteins, the biochemical workhorses of the cell. A single change in a single nucleotide can alter one of those amino acids; and if the change takes place in a key position, it can alter the function of the protein, sometimes dramatically. That's why it's appropriate to describe the code as digital.

Every time an *E. coli* divides, which can be as often as once every thirty minutes, the entire genome must be copied and the two copies carefully separated so that each daughter cell receives an identical set of genes. Sometimes this happens perfectly, but sometimes little mistakes are made and one or two of the nucleotides are miscopied. Sometimes the mistakes are more serious, and parts of the genome don't get copied at all while other parts are copied twice. Sometimes genes will be duplicated, moved from one place to another, or turned completely around. These alterations are known as "mutations," from the Latin *mutare,* which means "to change." They are heritable changes in DNA, in the genes, and they happen all the time.

Since the code is digital, even a tiny change, just one base in a genome of 4 million, has a chance to affect the characteristics of an organism. The replication mechanism means that most of the time that change will be transmitted faithfully to the next generation.

Now, if genes are nothing more than strings of letters from a four-letter alphabet, it follows that any letter of any gene can be changed, replaced, or shuffled by mutation. And so they can. Where's your mechanism? Right there. Mutations, acting upon that digital genome, produce variation, the raw material upon which natural selection goes to work. Since mutations can duplicate, rearrange, or change literally any gene, it follows that they can also produce any variation.

So, what's left to object to? Well, one could admit to the reality of the mutational mechanism, but then claim that the results of mutation are almost always bad. These undirected changes in the code tend to mess up the system, producing incorrect messages, malfunctioning proteins, and reducing an organism's fitness. Therefore, they aren't good candidates for a mechanism that depends on natural selection to find *increases* in fitness.

The kindest thing one can say about that argument is that it's simply, demonstrably, and clearly wrong.

Want proof? As we saw in the discussion of the evolution of drug-resistant bacteria (see Chapter 2), any physician in the infectious disease ward of a major hospital can cite chapter and verse on the problems posed by the success of this mechanism. In fact, the emergence of strain after strain of pathogen with resistance to widely used antibiotics is testament to just how quickly the mechanism of mutation and natural selection can produce beneficial mutations (beneficial, in this case, to the germ—not to us!).

It does no good to offer a counterargument that the genes for resistance to these antibiotics may have always been present in the population, for the very simple reason that brand-new resistance genes can be shown to emerge even under controlled laboratory conditions. In fact, if one actually believed that bacteria could not evolve genuinely new resistance genes, then all we'd have to do to achieve total victory against such diseases would be to produce a few brand-new, totally synthetic antibiotics. With no preexisting genes around, and with the evolution of new beneficial mutations impossible, a new, synthetic antibiotic should be invincible. Unfortunately, medical researchers know that this is just not true. Under the right conditions, evolution eventually produces and selects for resistance genes to every antibiotic, natural or synthetic.

Even viruses can do this—and as we have seen, some viruses, including HIV, are especially good at it.

Let's suppose that you are still skeptical, that you still want detailed, completely documented proof in a well-defined experimental system that the mechanism of evolution works, that it can produce and select for beneficial mutations. Fair enough.

I will present one more example, chosen not because it is so exceptional and dramatic, but rather because it is so mundane and routine, from a simple article published in a research journal of biochemistry.

Nearly all organisms, ourselves included, live in oxygen-rich environments. Most of the time this is a good thing, especially if you like to breathe. But oxygen, as important as it is to our metabolism, causes serious problems at the cellular level that biochemists have been aware of for years. Oxygen is highly reactive, and can interact with water and other cellular compounds to produce hydrogen peroxide (the stinging disinfectant used to sterilize cuts and abrasions) and an even more reactive chemical agent known as a hydroxyl radical (abbreviated $HO\cdot$). These reactive oxygen compounds can damage DNA, for which there are well-understood repair mechanisms, but they can also damage proteins. Once a protein is badly damaged by oxidation, the cell's only solution seems to be to junk the damaged macromolecule and make a new one.

A group of researchers, led by E.C.C. Lin at Harvard Medical School, wondered if they could learn more about what makes a protein susceptible to oxygen damage. They took a bacterial protein and set up conditions that might favor the evolution of a new protein resistant to such damage.

They chose a protein enzyme that was normally produced only when the organism was growing by fermentation in the absence of oxygen. Its actual name is *L-1,2-Propanediol:NAD⁺ 1-oxidoreductase,* but we'll call it ProNADO. Not surprisingly, since it's generally not around when oxygen is present, this protein is easily damaged by oxygen.

They grew these bacteria in the presence of oxygen but supplied propanediol, a small organic compound, as a source of food. The only way for this bacterium to use propanediol as a food source is to convert it to other compounds, and to do that, it must produce ProNADO, even though oxygen is present. Sure enough, in two separate experiments, mutants appeared in which the gene for ProNADO was switched on all the time, even in the presence of oxygen. Score one for evolution.

Now the tough work. Because ProNADO was not adapted to work in an oxygen-rich environment, it was easily damaged by the HO radical and other oxidizing compounds. The researchers themselves didn't know enough about oxygen damage to understand how to engineer a protein to make it resistant, but they figured they could let evolution teach them. Therefore, they simply grew the mutant bacteria in plenty of oxygen, figuring that sooner or later those chance, random mechanisms of mutation so scorned by critics of evolution would come up with an oxidation-resistant form of ProNADO.

Sure enough, mutation and natural selection did the trick. It produced two mutants, each of which was dramatically resistant to oxidative damage. Normally it takes about forty minutes of oxidation to completely inactivate ProNADO. In the first mutant, the enzyme retained forty percent of its activity after the same treatment; and in the second, nearly sixty percent. The researchers then analyzed the mutant genes and discovered that the changes were remarkably simple. In the first mutant, a single amino acid (the seventh in the chain) had been changed from an isoleucine to a leucine; and in the other, a leucine in the eighth position had been changed to a valine.

What might happen, the team wondered, if genetic recombination were to put both mutations together in the same gene? Bingo! Eighty percent resistance to oxidation! In fewer than two hundred bacterial generations, the mechanism of random, undirected mutation had taken the gene for an oxygen-sensitive protein produced only during fermentation and changed it into one that is switched on all the time and is highly resistant to oxygen damage.[23]

These results were not exactly earthshaking. They were routine, in fact, and that's exactly the point. The experiment, part of a solid experimental study, simply altered the growth conditions of a bacterium and then allowed evolution to produce a series of beneficial mutations to suit new conditions. In so doing, the researchers learned a series of valuable lessons on how to design proteins to make them more resistant to oxidation. That was what their paper and their research was really about.

What if they had taken a different route? What if they had decided to publish their results in a journal on evolution, claiming they had proof that evolution could redesign a gene in just a few hundred generations? I strongly suspect that the editors would have rejected it out of hand—not because of scientific flaws, but because their study lacked novelty. There simply is nothing new to the observation that the mechanism of evolution works.

Incredibly, the critics of evolution continue to claim that the mechanism of evolution is unknown, that mutations are never beneficial, and therefore that the hand of design is the only way to explain adaptations of organisms to their environments. For some reason, the real importance of simple studies like Lin's never seems to sink in—that natural selection is the driving force behind genetic adaptation, and that laboratories around the world have repeatedly observed the mechanism of evolution in action under controlled conditions.

Macro and Micro

You might think that scores of biochemical and molecular studies like Lin's work on the evolution of oxidation resistance would dispel any doubt that the mechanism of evolution has been demonstrated. Intelligent-design advocates like Phillip Johnson sometimes use a word game to pretend that such experiments mean nothing. When he and I discussed the issue of evolution on a PBS-sponsored Internet Web site, he dismissed such evidence with a disdainful flourish:

> The mutation/selection mechanism has never produced anything more impressive than variations in preexisting populations (microevolution).[24]

Curiously, his statement ignores the fact that such variations are exactly what natural selection works upon. Johnson then argues by manipulating the definitions of words. He defines these variations in a population as "microevolution." As the name implies, "micro" means small. Johnson is careful never to give this term an exact definition, to say exactly how small a change has to be to qualify, but nonetheless he applies the label to anything we can actually see or verify in the laboratory or field. Therefore, he would define the evolution of an oxygenation-resistant form of a protein as nothing more than microevolution.

Why is this significant? Johnson would have us believe that the only kind of evolution that really matters, the kind that really might produce new species, would have to be something different, something big, something macro. He wants to see macroevolution. And he claims that's exactly what Gould and Eldredge say the fossil record requires to support the rapid changes described by punctuated equilibrium.

Semantically, this is a brilliant strategy. You label any *observed* evolutionary mechanism as micromutational, say that yes, the work is interesting, but unfortunately the experiments at hand do not address the issue of macroevolution. By pretending that every example ever discovered of evolution in action is "just" microevolution, you can disqualify whole categories of important evidence against your case.

Johnson's argument is wrong on two counts. First, among bacteria, which is to say among most of the cells alive on the planet, it is particularly easy to see that the differences between species are indeed nothing more than the sum total of differences between their genes. If microevolution can redesign one gene in fewer than two hundred generations (which in this case took only thirteen days!), what principles of biochemistry or molecular biology would prevent it from redesigning dozens or hundreds of genes over a few weeks or months to produce a distinctly new species? There are no such principles, of course, which is the first reason why he is wrong.

The second reason is even more fundamental. We distinguish one animal species from another by its *morphology,* its appearance, the shape and size of its bones and body. Johnson argues that every appearance of a new species in the fossil record is a case of macroevolution, demanding an entirely new mechanism to produce such large changes.

Just how large, of course, he is careful not to say. If Johnson were a serious scientist, he might try to quantify the changes in morphology that have occurred in the fossil record, and then determine the rate of structural change that would be required to fit the observed speed of macroevolution. It may not surprise you that Johnson has not done this, but serious scientists have.

In 1981, biologist David Reznick started an interesting experiment. He scooped up a handful of guppies (*Poecilia reticulata*) from a waterfall pool in a stream in Trinidad, and transported them well upstream, beyond the waterfall, to a different pool. The lives of guppies, those unremarkable fish that grace the tanks of so many first-time aquarium owners, may seem relaxed and stress-free, but the natural habitat of this fish fits every bit the description of nature red in tooth and claw. Bloodthirsty predators, most notably a vicious little cichlid fish, exact a terrible toll from the adult population, catching and eating the largest of the guppies.

The new location into which the researchers placed a sample of the population differed from the guppies' original home in just one important respect—it lacked the large predator. For eleven years they studied the sizes and reproductive patterns of the transplanted guppies. To be doubly certain that any changes they saw were genetic, they captured guppies from the wild, brought them back to laboratory conditions, and made all of their measurements on the second generation of laboratory-reared fish.

They found that the fish quickly adapted to their new environments. Fish bred from the upstream fish took longer to reach sexual maturity, and were larger at maturity than the downstream fish. These adaptations, of course, make perfect sense. With the cichlid preying on the largest fish, only those that reach sexual maturity early, when they are small enough to evade the large predator, would be successful in reproducing. Once the guppies were transplanted upstream to a predator-free environment, natural selection would give any tendency towards a longer period of growth before sexual maturity a big reward—the chance to produce more offspring, as the number of eggs a female can produce goes up with an increase in body size.

In one sense, Reznick's results are the unremarkable outcome of differing conditions of natural selection in two habitats. But the care of their study enabled them to do something else—they measured the rate at which these heritable, genetic changes took place.

Changes in any measurable, heritable characteristic can be expressed in terms of units known as "darwins." A rate of change of one darwin means that the average value of a measurable feature has changed by a factor of 2.718 over the course of 1 million years.[25] Reznick and his associates carried out two sets of experiments in different streams, and measured rates of change for various characteristics from 3,700 to 45,000 darwins. How does this compare to historical rates of change observed in the fossil record—rates that Johnson regards as requiring a special macroevolutionary process? The changes in tooth size among horses during the Tertiary, a period of rapid (in evolutionary terms) evolution well recorded in the fossil record, is around 0.04 darwins. Much earlier, the record documents an even more dramatic increase in size from the smallest to the largest of the ceratopsian dinosaurs, a lineage that includes *Triceratops,* that is widely regarded as an example of rapid evolutionary change. Here the change is faster: 0.06 darwins!

If you notice something remarkable here, join the club. The authors picked it up right away. They wrote that the rates of change they observed "are similar in magnitude to rates that have been obtained by artificial selection and four to seven orders of magnitude greater than those observed in the fossil record."[26] In other words, these experiments carried out under field conditions produced rates of genetic change that were 10,000 times to 10,000,000 times faster than the typical rates of rapid genetic change observed in the fossil record.

No one would seriously argue, even for a moment, that novel genetic mechanisms were needed to account for the effects of these eleven years (about eighteen generations) of natural selection on Reznick's guppies. Why then would anyone find it necessary to demand novel mechanisms to account for the much, much slower change observed over geological time? Remembering that the theorizers of punctuated equilibrium had wondered out loud if new genetic mechanisms were necessary to sustain the relatively rapid changes they observed in the fossil record, Reznick and his associates hit this theoretical nail right on the head:

> Specifically, Gould and Eldredge argue for the necessity of bursts of speciation followed by species selection to sustain the rapid change associated with punctuations in the fossil record. Our work cannot address the efficacy of mechanisms other than natural selection, but it extends our understanding of what is attainable through this process. It is part

> of a growing body of evidence that the rate and patterns of change attainable through natural selection are sufficient to account for the patterns observed in the fossil record.[27]

Bingo. And these results were not an isolated case. Recent studies involving finches, lizards, and even the common house sparrow[28] document rates of genetic change of 300 or 400 darwins, easily 1,000 times faster than the supposedly fast-paced macroevolutionary events demanded by punctuated equilibrium.

What this means, of course, is that the microevolutionary processes so scorned by critics of evolution are more than sufficient to account for even the fastest transitions documented in the fossil record. The mechanism isn't missing, it isn't too slow, and it isn't even theoretical. The mechanism of evolution is real, is observable, and is more than adequate to the task at hand.

On the Joys of Being Outrageous

Against the backdrop of laboratory experiment and field observation, the demands for a mechanism made by latter-day creationists like Johnson collapse into what has been called the "argument from personal incredulity." The only compelling case they can make against evolutionary theory is that the mountain of historical and experimental evidence supporting evolution hasn't yet convinced *them*. In scientific terms, the hollow nature of such contentions speaks for itself, and is probably the principal reason why such anti-evolutionists avoid scientific meetings—where they would face informed critics of their ideas.

Nonetheless, the opponents of evolution are certainly entitled to ask, if the ordinary mechanisms of mutation and natural selection are more than sufficient to account for the patterns of change observed in the fossil record, then why do some of the most prominent evolutionary biologists seem to say that this is not the case? Specifically, why have the likes of Gould and Eldredge written that the fossil record is distinctly non-Darwinian, and why have they argued that a new theory—punctuated equilibrium—is required to explain it?

The answer is surprisingly mundane. The driving force behind punctuated equilibrium was an attempt to enhance the novelty and importance of a relatively modest observation about the fossil record.

A *modest* observation? But haven't we already established that the work of Gould and Eldredge presented a dramatic alternative to the dogmatic phyletic gradualism advanced by Darwin? Not really. And this is the point that bears closer inspection.

Let's begin by taking another look at Figure 4.2, taken from a 1980 drawing to highlight the differences between punk eek and phyletic gradualism. At first glance the two diagrams look quite different. In one the change is sudden, and in the other the change is gradual and continuous. But look closer and you'll see that something fundamental is missing from these diagrams—a scale! To be sure, the Y-axis is labeled "Time" and the X is "Morphology." But how much time and how great a change in morphology? The architects of the theory don't say—and they have never said.

A little tinkering shows why this is important. Suppose that we examine punctuated equilibrium by zooming in on that relatively brief period of time in which a new species is established (see Figure 4.5).

If we take that period from the punk eek drawing, and expand those arbitrary and unspecified time units, what happens? The drawing of this so-called revolutionary theory of change suddenly becomes transformed into a model of the very gradualist pattern against which it was contrasted. In other words, the only thing that makes punctuated equilibrium appear novel, or even different from gradualism, is the choice of a timescale with which to present the data. And since a timescale is not presented as part of the theory, there really becomes no way to distinguish what kind of data support gradualism and which might support punctuation.

It sometimes is argued that the real novelty of punctuated equilibrium can be found in its recognition of the unevenness of evolutionary change. Darwinian change is then portrayed as requiring that such change be slow, steady, and continuous. But that's not at all what Darwin said. Remember that he drew exactly one diagram in *The Origin* (see Figure 4.6 for close-up, Figure 2.4 for complete diagram), designed to allow his readers to see how he believed new species were formed over time.

Darwin presented this diagram so that his readers would have some idea of how he imagined descent with modification might take place over time. Naturally, he focused on what things might look like when changes were taking place—not on what might happen when they were not. Should we take the branching lineages that Darwin drew as show-

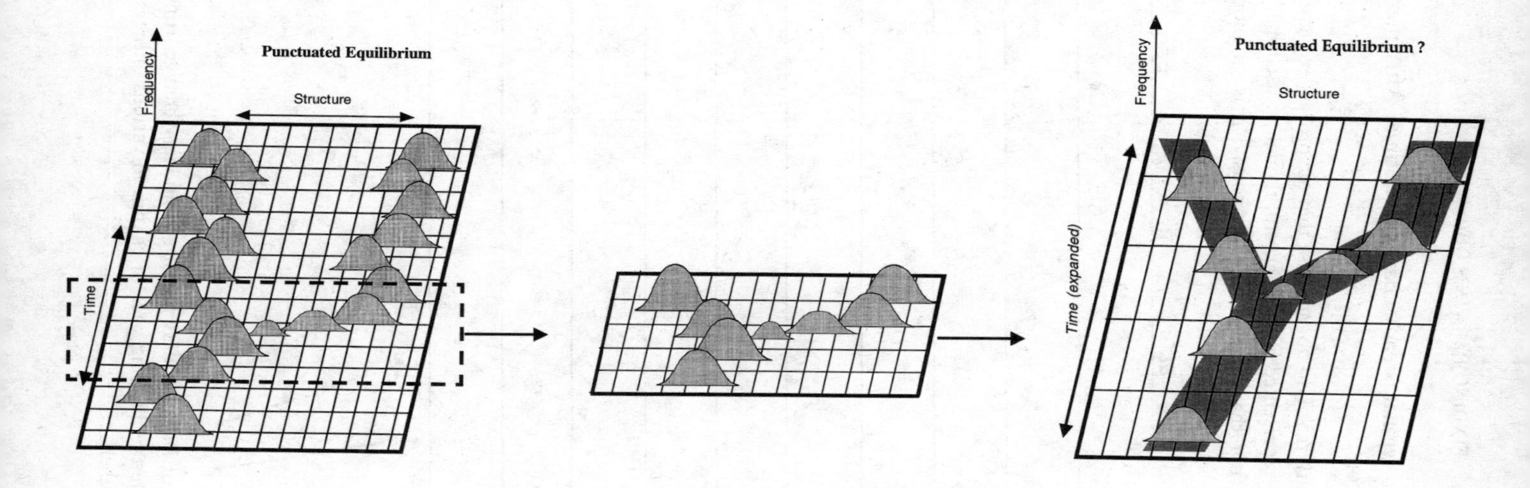

Figure 4.5. The scale of time is crucial in assessing whether the "rapid" bursts of speciation associated with punctuated equilibrium are distinctly different from those associated with phyletic gradualism. If, as shown here, the crucial part of a classic representation of punctuated equilibrium (left) is redrawn with its arbitrary timescale expanded, the result (right) is a diagram of evolutionary change nearly identical to phyletic gradualism.

From Figure 4.2, which is redrawn from R. Lewin, "Evolutionary Theory under Fire," *Science* 210 (1980): 883–885.

ing his commitment to regular, steady change over time? After presenting the drawing he wrote:

> But I must here remark that I do not suppose that the process ever goes on so regularly as is represented in the diagram, though in itself made somewhat irregular, nor that it goes on continuously; it is far more

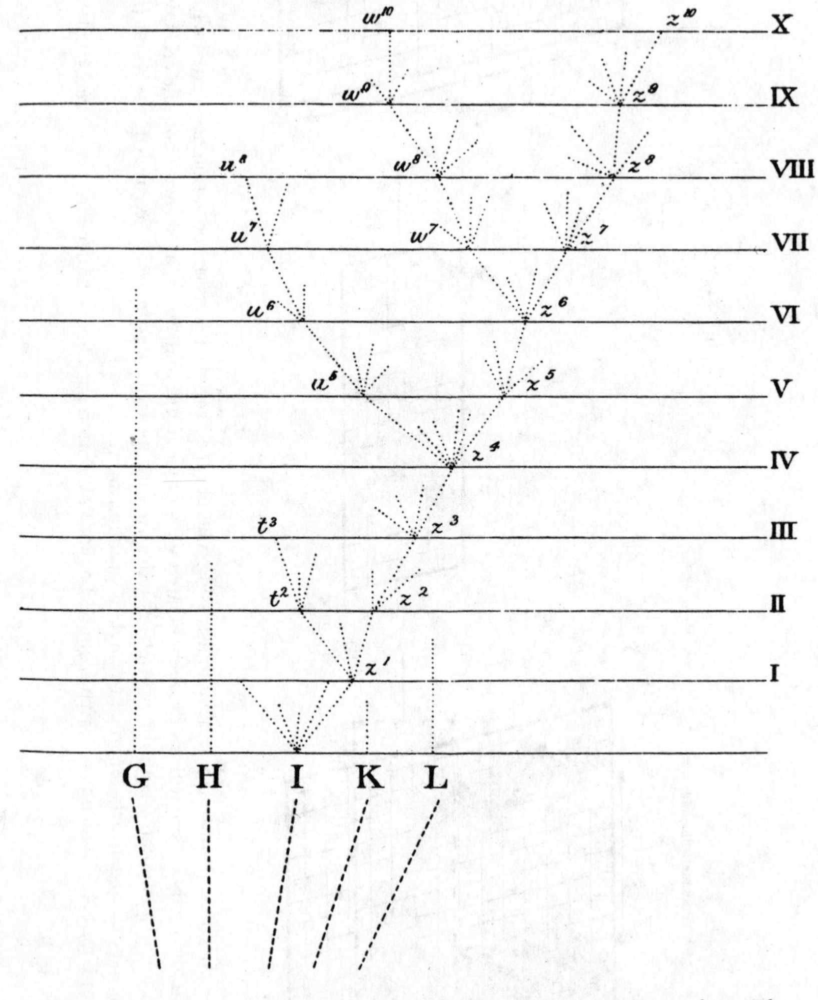

Figure 4.6. A portion of Darwin's drawing of evolutionary change from *The Origin of Species*. Note the differing pace of change in various lineages, and the lack of change in others.

> probable that each form remains for long periods unaltered, and then again undergoes modification.[29]

Can we even conceive of a better definition of stasis, or equilibrium, than the notion that a species "remains for long periods unaltered?" Or of a better definition of punctuation than the species "again undergoes modification?" A visitor to the land of evolution-speak might be forgiven for coming to the conclusion that the controversy over gradualism versus punctuated equilibrium was a wee bit contrived. And so it was.

Here we have pinpointed the essential secret of punctuated equilibrium. Wouldn't it have been wonderful to look at the range and diversity of life, and to draw from one's observations a sweeping and transforming generalization that changed everything we thought we understood about natural history? I have no doubt that's exactly what Eldredge and Gould sincerely believed they were doing when they first described their theory. For better or for worse, that's also the feat that Darwin accomplished; and unfortunately for my colleagues, it turns out that Darwin got it positively and overwhelmingly right the first time.

To be sure, there is plenty of action and excitement in the field of evolutionary biology. Much of this comes from the new technical capabilities of molecular genetics, which have just begun to make it possible to study the evolutionary dynamics of individual genes and even whole genomes. On the point of fashioning, as Darwin did, a broad and general theory that would revolutionize our understanding of natural history, the proponents of punctuated equilibrium have simply missed the boat.

Why does this matter? It makes a difference because of the way in which Phillip Johnson and other critics of evolution have pounced upon such words and phrases as "instantaneous" in the punctuated equilibrium description of the origin of species. To biologists, punctuated equilibrium is part of an interesting but highly technical argument about whether major evolutionary trends in size and structure are accomplished by the production of new species or by the transformation of existing species. In either case, there is plenty of evidence that the transformations themselves have taken place repeatedly in natural history, so why should a creationist care about the pace of such changes? For Johnson, reading about "instantaneous" origins for natural species opens the door to the miraculous. In his mind, only the materialist prejudices of mainstream scientists constrain them to slam that door shut.

The Emperor's New Theory

Ironically, the appeal Gould and Eldredge hold for creationists comes from their acknowledged status as evolutionary biologists of the first rank. This lends an air of authority to their generalizations about the nature of the fossil record. Presumably, if *they* say most species appear suddenly and without apparent ancestors, then surely it is so. And just as surely, these generalizations must mean that Darwinian evolution—against which Gould and Eldredge seem to argue—cannot explain natural history.

This line of reasoning is a clear case of hearing only what you want to hear. The real point at issue is a simple one. What actually happens during those "instantaneous" transitions to new species? The hope of anti-evolutionists like Johnson, of course, is that something special, something non-Darwinian, something truly miraculous takes place during each of those horizontal leaps. Searching the record of the past (or perhaps more accurately, searching the writings of evolutionists) for the magic of creation, they hope that every one of those "punctuations" made so famous by Gould might be evidence of the Creator at work.

Well, are they? Does the evidence allow us to say that the Creator might be working His magic at each of these punctuational events? Putting it another way, what would we see if we could observe the origin of a new species directly, in real time? Would we see the sudden appearance of a radically new species out of thin air? Or would we see classic series of intermediate forms? On this score Gould and Eldredge have been uncharacteristically clear. First, they took pains to point out they had not claimed punctuated equilibrium to be

> a new discovery, but only a novel interpretation for the oldest and most robust of paleontological observations: the geologically instantaneous origination and subsequent stability (often for millions of years) of paleontological "morphospecies."[30]

The key to understanding this statement is the term "geologically instantaneous," words that the creationists conveniently gloss over when invoking the punctuational description of the fossil record. What, exactly, does this mean? Once again, Gould and Eldredge have made this clear:

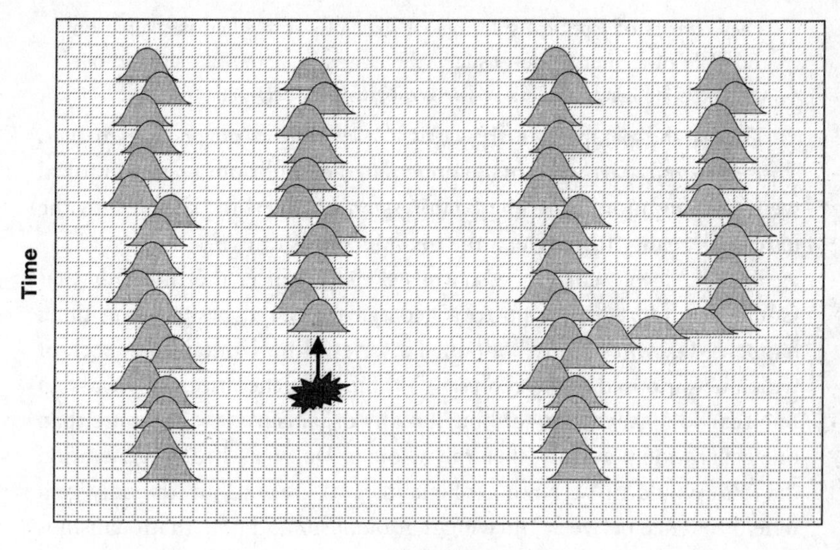

Figure 4.7. Two views of punctuated equilibrium. The anti-evolutionist advocates of intelligent design maintain that the rapid origin of species might be evidence for special creation of new species (left). Evolution, however, would maintain that the real reason for the apparently instantaneous origin of a new species from a closely related one is the compression of real time into geological time (right).

> We realized that a standard biological account, Mayr's peripatric theory of speciation in small populations peripherally isolated from a parental stock, would yield stasis and punctuation when properly scaled into the vastness of geological time—for small populations speciating away from a central mass in tens or hundreds of thousands of years, will translate in almost every geological circumstance as a punctution on a bedding plane, not gradual change up a hill of sediment, whereas stasis should characterize the long and recoverable history of successful central populations.[31]

Fair enough. This statement means that appearance of instantanous speciation is really just an artifact of the compression of time that takes place each time a new layer is formed in the fossil record. Another of Gould's essays makes this point abundantly clear:

> These events of [species] splitting are glacially slow when measured on the scale of a human life—usually thousands of years. But slow in our terms can be instantaneous in the geological perspective.[32]

This is the perspective the critics of evolution never seem to get. Those fits and starts that made punctuated equilibrium seem so radical were nothing more than the compression of Darwin's sketch into the hard rock reality from which the fossil record is drawn.

In fact, if we "correct" Darwin's sketch to meet the timescale that Gould and Eldredge assign to a geological instant, a surprising thing happens. Darwin's figure contained a series of horizontal lines, each of which he supposed might represent a thousand generations. If we assign one year to a generation (although many organisms have far more than one generation per year), then the whole of Darwin's 14,000-generation diagram would fit within a geologic instant! Once again, the only genuine difference between Darwinian gradualism and punctuationalism is an arbitrary choice of the timescale over which to display the data. Once this is evident, even its originators' claim of a "novel interpretation" wears a little thin.

The real test, however, of whether or not I have presented a fair description of the lack of theoretical novelty in punctuated equilibrium is whether or not a perfectly preserved set of specimens from one of those instantaneous transitions would show a true series of intermediates of the sort predicted a century and a half ago by Darwin. In one of those ironic little twists of scientific fate, exactly such specimens found their way, not very long ago, right into the hands of Stephen Jay Gould.

Much of Gould's own field work has involved studies of the Bahamian land snail, *Cerion*. As a landlocked electron microscopist whose research work has never taken him to any location more exotic than the university's greenhouse, I applaud Gould's informed choice of model organisms. But *Cerion* offers far more than a nice place to travel to. The snail's plasticity has given rise to a number of new species in geologically recent times. The ease with which its shells fossilize means that many of these speciation events can be studied in detail.

As Gould tells the story, in 1990 he and David Woodruff (of the University of California, San Diego) collected a remarkable series of shells on the island of Great Inagua. Today, this island is home to *Cerion rubicundum,* a snail whose distinctive ribby shells are particularly easy to identify.

Fossil deposits, even very recent ones, from Great Inagua lack *C. rubicundum* entirely, and instead contain the largest of all *Cerion* species, *C. excelsior*. Gould and Woodruff believed that the transition, over time, had been accomplished by species hybridization, with the average morphology of the snails on the island changing gradually from *C. excelsior* to *C. rubicundum*. They even collected a series of shells that, in Gould's words, "seemed to span (and quite smoothly) the entire range of form from extinct *C. excelsior* to modern *C. rubicundum*." All of the shells had been collected in a single mud flat, the equivalent of a single bedding plane in a geologic column. This meant that their relative ages could not be determined, and therefore there was no way to know whether or not the smooth transition between species had actually taken place as they imagined.

For years, Gould's specimens sat in a museum drawer. Then geochemist Glenn Goodfriend learned of Gould's specimens, and the two of them teamed up to put the shell fossils into their correct chronological order. By combining radiocarbon dating with another technique that measures long-term changes in the amino acid composition of the shell, the two scientists were able to put their specimens into the proper order. The result was impressive. As shown in Figure 4.8, spread out over 15,000 years, the size and shape of the older shells changed smoothly into the modern shape of *C. rubicundum*.

Cerion might be just a snail, but the good luck of finding so many specimens from a single geologic period provides perfect documentation of what a species transformation actually looks like under the microscope of time. And it's not at all what the critics of evolution would have hoped for. A modern species doesn't come out of thin air, but arises by descent with modification of an ancient ancestor. Darwin examined, Darwin confirmed.

As Gould and Goodfriend noted:

> In favorable circumstances such as these, the fossil record can supply direct evidence for evolutionary change at the generational scales of microevolutionary processes, rather than the conventional paleontological scale of general trends over millions of years.[33]

An observer might be forgiven for noting the obvious, which is that when one of those instantaneous blips described by punctuated equilibrium is examined up close, it doesn't look much like punctuation at all,

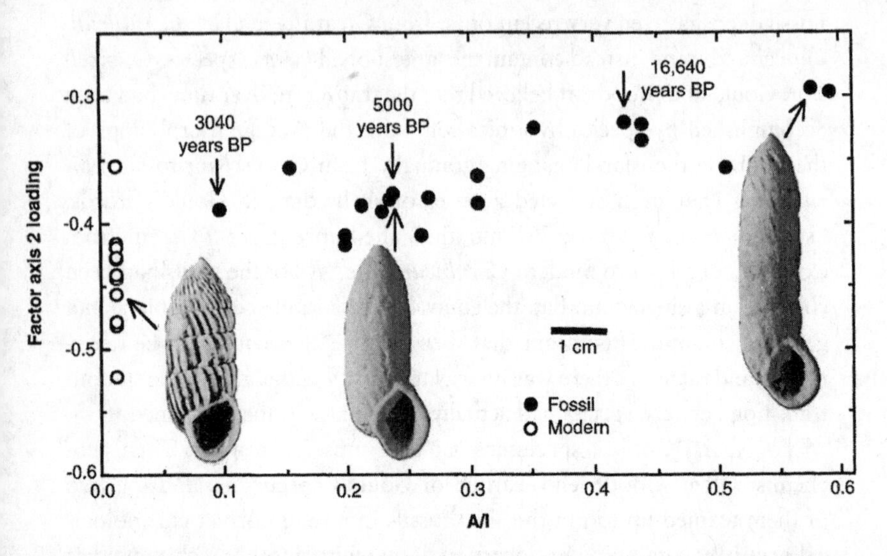

Figure 4.8. Changes in shell morphology of fossil and modern *Cerion* shells. The nearly complete fossil record shows a smooth, continuous transformation from the ancient species to the modern one. The horizontal axis represents time ("A/I" values measure changes in amino acid composition, which increase with time. Several exact ages are indicated), and the vertical axis represents a combined measurement of several variables in shell morphology.

From a 1996 study by Goodfriend and Gould, *Science* 274 (1996): 1894–1897.

not even when Steve Gould himself is doing the examining. It looks instead very much like Darwin's gradualism.

So, how did Gould react to the obvious implications of his own work? In his monthly column in *Natural History,* he recounted his discussions with a reporter from *Science* magazine:

> He wanted to write an accompanying news story about how I had found an exception to my own theory of punctuated equilibrium—an insensibly gradual change over 10,000 to 20,000 years. I told him that, although exceptions abound, this case does not lie among them but actually represents a strong confirmation of punctuated equilibrium. We found all 20,000 years' worth of snails on a single mud flat—that is, on what would have become a single bedding plane in the geological record. Our entire transition occurred in a geological moment and rep-

> resented a punctuation, not a gradual sequence, of fossils. . . . The reporter, to his credit, completely revised his originally intended theme and published an excellent account.[34]

The reporter should have stayed with that original story, because his first instincts were overwhelmingly correct. Gould had *indeed* found an exception to the way in which punctuated equilibrium is generally presented, explained, and understood. Naturally, he resisted the conclusion that this experimental evidence contradicted his well-known theory, and did his best to explain it away. In saving that theory from the gradualist dangers of his own *Cerion* data, he showed just how little is left to distinguish punctuated equilibrium from Darwin's orginal formulation of the pace of evolutionary change. Namely, to repeat Darwin's own words, that *"each form remains for long periods unaltered, and then again undergoes modification."* To this, all that punk eek really has to add is the fact that the compression of geological time will make those periods of modification seem sudden and abrupt.

To be fair, there are a few interesting technical elements to punctuated equilibrium that I have not mentioned, most notably the assertion that major evolutionary trends are strongly correlated with the production of new species. For us, the point at issue is simpler: Do new species appear so suddenly that they require a mechanism above and beyond the ordinary processes known to take place in genetics and molecular biology? And the answer is simple, too. They don't.

Case Considered

I roomed with a law student my first year in grad school at the University of Colorado, and absorbed a few stray lessons as he, like me, struggled to learn his craft. One of those was a valuable answer to the question, When do you really need a lawyer? The best answer was, according to my roommate, Whenever the other guy has one. My friend's logic was unassailable. Lawyers are rhetorical gunslingers, professional advocates; love them or hate them, lawyers are people who argue for a living, and if you're going up against a lawyer on your own, you are in trouble.

Phillip Johnson is a lawyer, and a good one. The rhetorical and logical skills he brings to any argument are part of the necessary training for the proper practice of law. But making a case in law is not the same thing as making a case in science.

Lawyers are advocates. Their task is not necessarily to seek the truth, but to present a particular point of view as forcefully and effectively as they can. The whole structure of the American judicial system relies on the presence of two opposing advocates—the two sides of a case—whose arguments will, under ideal circumstances, contribute to the search for truth. A judge has the responsibility to make sure the advocates play by the rules, and another group—the jury—weighs both sides and makes the ultimate decisions of fact. The law profession produces people who are skilled in the preparation and presentation of argument, and that certainly describes Johnson.

Science is different from the law. I say this not because I am naive enough to believe that scientists are always fair and lawyers are always biased, or because scientists seek the truth while lawyers seek advantage. Like law, science involves a process, and the pursuit of scientific truth is hobbled with all the problems, weaknesses, and frailties that attend the human condition. Scientists lie, cheat, and steal, and they make mistakes. Plenty of mistakes. But science corrects those mistakes, and the mechanism of their correction cuts to the heart of the difference between science and law.

In law, there is no perfect solution, no absolute justice, no single and unique standard to balance rights and responsibilities. The whole practice of law in a democratic society is predicated on the notion that the people, through their elected representatives, may set and change the accepted norms of legal behavior, and even the rules by which the legal profession operates. In such a system, the role of advocate is enhanced, especially if the advocate is able to influence the making of rules themselves.

In science, there is an absolute, unchanging authority. That authority is nature. And there really is just one perfect solution for any scientific question, namely, the one that corresponds to actual, physical reality. It is fashionable in some academic circles to pretend that science names nature as the ultimate authority only to disguise its true allegiance to the white, male, Western power structure. And it certainly is true that the practice of science, as a human institution, is affected by the lines of authority within society. But science transcends such arguments for the very simple reason that it really does depend, as its final judge of truth, on the objective reality of the world around it. This doesn't mean that science always gets to that reality right away, and it certainly doesn't mean that every pronouncement of the scientific establishment, including the ones about evolution, should be taken as revealed truth. It does mean that one who would advance a scientific theory has a task quite different from that of a legal advocate.

When I first read Phillip Johnson's book, *Darwin on Trial,* I read it as a scientist, and it puzzled me. In every chapter he attacked what he considered to be a weak spot of evolutionary theory, implying in each and every case that there might be another explanation, a better one than evolution. This is a common strategy in scientific argument. As I neared the end of the book, I expected Johnson to do what any one of my scientific colleagues would do at the conclusion of a provocative seminar—to lift the curtain and reveal that better explanation. Like any scientist, I expected him to present a model that would fit the data more precisely, a model that would possess powers of explanation and prediction well beyond the theory he had attacked. But Johnson did nothing of the kind.

Gradually I realized that the case he and his associates bring against evolution is *not* a scientific case at all, but a legal brief. The goal of his brief is to raise reasonable doubt, to create a climate in which the intellectual claims of evolution seem shaky, even unreasonable. What it never does is to present us with an alternative—any alternative—to the seamless integration of theory and natural history provided by evolution. There are two reasons for this.

First, the only message of Johnson's book as a legal brief is that case for evolution is weak. All he needs to establish is that the other side has not proven its case beyond a reasonable doubt. A skilled advocate, he sees no need to present a case of his own, and refuses to do so even when asked.[35]

The second reason is tactical. From time to time, Johnson allows that there just might be an alternative to evolution, an alternative known as "design." Fair enough. To lift the curtain on this supposed theory, we'd have to ask what design might look like to an observer. What would it mean in terms of the sequential character of the fossil record? How does design explain the appearances and extinctions, for example, of twenty-two different elephant-like species in just the last 6 million years?

As we have already seen, design can be joined to the facts of natural history only in a way that makes the designer look like a comic, overworked, and rather slipshod magician. Always a good lawyer, Johnson knows that if he begins to explain the operation of this theory, the case against evolution will evaporate. But his goal, after all, isn't science. It is to undermine a scientific theory the implications of which he does not like. So he avoids mention of any alternative theory, and works at establishing reasonable doubt.

What's the proper response? Quite a few scientists believe that such attacks on evolution are best ignored, that they aren't worthy of a response. I don't agree. And I don't think that this is a debate in which science should call in lawyers of its own. As complex as science sometimes seems, the scientific resolution of dispute and controversy is actually much simpler than it is in law or philosophy. The best way to answer any objections to evolution is see whether they are grounded in scientific fact.

When that is done, Johnson's objections collapse. His claim of a missing mechanism is easily refuted, his hopeful misinterpretations of punctuated equilibrium fall apart under close scrutiny, and his assertion that the fossil record does not support evolution is in error.

In *Darwin on Trial,* Johnson disputes many of the details of the evolutionary record of the vertebrates. The facts of that record show that in the Upper Devonian period, about 350 million years ago, there was a group of freshwater fish known as rhipidistians generally thought to be ancestral to the tetrapods, four-footed land vertebrates. Johnson notes that rhipidistian fish are similar to modern lobe-finned fish such as lungfish, and then claims that the lobe-fins give "no indication of how it might be possible for a fish to become an amphibian," suggesting that "a rhipidistian fish might be equally disappointing to Darwinists if its soft body parts could be examined."[36] The implication is that tetrapods didn't really evolve from fishes or, for that matter, from anything else.

A career in the law could not have prepared Phillip Johnson for the pace of discovery in science. Two important fossil finds, one on each side of the fish-to-amphibian transition, have crushed his argument. In 1991, two British scientists reported on an unusually detailed skeleton of *Acanthostega gunnari,* a remarkably fish-like tetrapod. Paleontologists have recognized for years that the earliest tetrapods retained scores of fish-like characters, but this specimen of *Acanthostega* was so well preserved that it contained a genuine surprise—internal gills. No other amphibian possesses internal gills, and the structures preserved within the fossil make it clear that *Acanthostega* could breathe with its gills underwater, just like a fish, and could also breathe on land, using lungs. It was a true transitional form. This first amphibian-like tetrapod was, as evolution would have predicted, more fish-like than any tetrapod to follow. As noted, the authors summarized the importance of their find by writing that "Retention of fish-like internal gills by a tetrapod blurs the traditional distinction between tetrapods and fishes."[37]

It gets even better. In 1998, an opportunity to examine the soft parts, as Johnson put it, of a Devonian fish surfaced along a roadside in Pennsylvania. Paleontologists Edward B. Daeschler and Neil Shubin discovered a fossilized fin so well preserved that its soft parts could be seen outside its underlying bony skeleton. The fin contained eight well-defined, recognizable digits. Incredibly, this fish had a fin with fingers, eight in number, just like the digits of the earliest tetrapods. Hardly the disappointment to Darwinists Johnson had predicted!

So, what is the result of a hearing on this portion of our advocate's case? Not only were the first tetrapod amphibians remarkably fish-like and rhipidistian fish remarkably amphibian-like, but their mix of characters removes any doubts as to when and where the key tetrapod characteristics of breathing and locomotion first evolved. Every objection of Johnson's has been answered. The fossils have been found in exactly the right place, at exactly the right time, with exactly the right characteristics to document evolution.

Is this just a tactical victory for evolution, one isolated success in a series of logical battles that could go either way? No. Remember the scientific stakes. If evolution is genuinely wrong, then we should not be able to find *any* examples of evolutionary change *anywhere* in the fossil record. A desperate advocate could quibble, of course, that the evolution of the modern elephant from its ancestor *Primelephas* is too small a change to matter. On this point, anyone who has seen these skeletons would surely disagree. The scale of the transformation from fish to tetrapod amphibian, however, is just too large to dismiss. It breaks the case of the defense.

More important, the implication that we should put Darwin on trial overlooks the fact that Darwinism has *always* been on trial within the scientific community. Once we invoke Darwin's mechanism, we immediately place testable constraints on the patterns of natural history. If organisms show common ancestry, then we should find a nested series of relationships among existing organisms, which in fact we do. We should find that novel organs and structures are found only in the actual descendants of ancient species in which those structures first appeared—also true. Finally, and what is most important, we should find a consistent pattern of ancestor-descendant relationships that expands as new discoveries fill in the details of the fossil record. The new fossils that document tetrapod origins are just two examples of the spectacular way in which this final test continues to be met by evolution.

To pretend that evolution has never been tested, or never been put on trial, is to overlook one of the strongest, more vigorous, and even the most contentious branches of modern science. The most effective scientist, unlike an effective attorney, is not the one who makes the best case for his side of a scientific issue. Rather, it is the one who finds the best way of testing every relevant idea against the objective reality of nature. When Phillip Johnson's case, well argued as it may be, is tested against this reality, it fails, and it fails every time.

A Theology of Magic

There's an old saying that the devil is in the details. If this is true of anything, it's certainly true when it comes to invoking design to explain the fossil record. At first glance, design seems to explain everything, because the principle of an all-powerful designer can be used to account for anything. This is one of its main frustrations to science. By definition, design cannot be tested, cannot be disproven, cannot even be investigated. The arguments for design are entirely negative in nature, and the writings of Johnson, Berlinski, and others confirm this in briefs that assert only the insufficiency of the evolutionary mechanism. Evolution doesn't work, they say, so the only alternative is design.

Would that it were that simple. Evolution does work, of course, and that is how I have answered their criticisms. What I find particularly distressing is that these anti-evolutionists fail to see the damage they have done philosophically and theologically by invoking design as a universal alternative to evolution.

To an anti-evolutionist, when new species appear in the fossil record, it is not as the modified descendants of ancestral species. It must be the product of design. This makes the fossil record a tool for recording the actions of the designer. We must say, then, that it pleased the designer to design only microorganisms for nearly 2 billion years of earth's history. He then began to tinker with multicellular organisms, producing a bewildering variety of organisms that survived only briefly. In the Cambrian era, roughly 530 million years ago, the designer produced an extraordinary variety of multicellular organisms, many of which were the first representatives of what we now regard as the animal phyla, the major groups into which animals are classified. Even in the Cambrian, he was not yet interested in designing a vertebrate, an animal that, like us, is built around a backbone. That came later.[38]

Then, as we have seen, the designer produced one organism after another in places and in sequences that would later be misinterpreted as evolution by one of his creatures. And just to compound that misinterpretation, he would ensure that the very first limbs he designed looked just like modified fins, and that the first jaws he designed looked like modified gill arches. He would further ensure that the first tetrapods had tail fins, like fish, and that the first birds had teeth, like reptiles. So thoughtful was this designer that after having designed mammals to live exclusively on the land, he would redesign a few, like whales and dolphins, to live in water—but not before he designed creatures that were literally halfway between land and swimming mammals. In working his magic, this designer chose to create forms truly intermediate between walking and swimming mammals.

It would be nice to pretend that this description is nothing more than an irreverent polemic, a nasty poke at the opposition. But it's not. It is a fair description of just a tiny bit of what any advocate of intelligent design *must* believe in order to square such beliefs with the facts of geological history. And this is just the beginning.

Intelligent design advocates have to account for patterns in the designer's work that clearly give the appearance of evolution. Is the designer being deceptive? Is there a reason why he can't get it right the first time? Is the designer, despite all his powers, a slow learner? He must be clever enough to design an African elephant, but apparently not so clever that he can do it the first time. Therefore we find the fossils of a couple dozen extinct almost-elephants over the last few million years. What are these failed experiments, and why does this master designer need to drive so many of his masterpieces to extinction?

Intelligent design does a terrible disservice to God by casting him as a magician who periodically creates and creates and then creates again throughout the geologic ages. Those who believe that the sole purpose of the Creator was the production of the human species must answer a simple question—not because I have asked it, but because it is demanded by natural history itself. Why did this magician, in order to produce the contemporary world, find it necessary to create and destroy creatures, habitats, and ecosystems millions of times over?

I have no doubt that some will read these questions as blasphemous. How dare I (or anyone) question the motives of God? But keep in mind who is really responsible for dragging the designer into the crucible of science. If we wish to make sense of natural history by invoking the miraculous to account for life's every twist and turn—and that is exactly what Johnson wishes to do—then it is the invoker himself who has presented in court the motives of the Creator. It is the advocate of design who wants to hold his designer responsible for every detail of life. Ironically he's the one, not the evolutionist, who has made these questions necessary.

In the final analysis, God is not a magician who works cheap tricks. Rather, His magic lies in the fabric of the universe itself. The fossil record is not a series of sequential tricks fabricated for no purpose other than to mislead. The fossil record represents, with all of its imperfections, an epic of evolutionary change, the history of life on this planet grand in its range and diversity and magnificent in its detail. It is the record of the historical process that led to us. It is the real thing, and so are we.

5

GOD THE MECHANIC

Science loves an underdog, and like any teacher of science, I've got a slew of underdog stories that I relate to my students. They include people like Howard Temin, who used the existence of RNA viruses to predict that there just *had* to be an enzyme capable of copying information from RNA into DNA. Temin figured that if a virus carried its genetic information in the form of RNA, but could still manage to subvert the DNA-controlled genetic machinery of a living cell, then the virus just had to have a way to copy its RNA-coded genes into DNA. He was so sure of this that he even proposed a name for his hypothetical DNA-making enzyme: "reverse transcriptase," because it would have to be able to copy the "transcript" of genetic information in a direction that was the *reverse* of its usual pathway. Not everyone thought Temin was foolish, and quite a few admired his boldness, but many, if not most, molecular biologists expected him to be wrong.

He wasn't. When an actual reverse transcriptase enzyme was discovered, his foresight was rewarded with a Nobel Prize.[1] Fortunately, Temin's willingness to take a chance to advance science is anything but unique. My own gallery of underdog heroes includes Barbara McClintock, whose decades-long studies of corn genetics showed that genes can "jump" from one position to another on the chromosome; Stanley Pruisner, who may (or may not)[2] have shown that infectious protein particles are capable of causing disorders like mad cow disease; and Thomas Cech, whose graduate students Paula Grabowski and Arthur Zaug at the University of Colorado showed that enzymes need not be made from proteins—that under the right conditions, RNA could act as an enzyme. Each of these discoveries was revolutionary, each directly contradicted prevailing wisdom in the field, and in each case the work of the underdog pushed science along to a new understanding of the workings of nature.

True, science can become political, scientific institutions can fail to recognize the value of novelty, and individual scientists may sometimes try to stifle critical dissent—but all of this is just another way of saying that science is a human activity. A glorification of the individual as iconoclast is part of the culture of science itself. It may even be one of the reasons why America, which so celebrates the culture of individualism, has proven such fertile territory for the scientific enterprise.

Does this mean that our current ideas about evolution might be modified, updated, or even thrown out altogether by a crusading underdog? Absolutely. Evolution is an exceptionally important scientific idea, but it is still just a scientific idea—subject to testing, possible disproof, and even replacement by a new and superior scientific theory. For that to happen, we'd need more than just a lone crusader. We'd need a crusader against evolution who can marshal the evidence to show that he or she is right.

The latest would-be underdog is Michael Behe, a professor of biochemistry at Lehigh University in Pennsylvania. In the minds of many opponents of evolution, Behe has taken up the glorious banner of intellectual revolutionary to correct an outmoded scientific idea, one that was born in the nineteenth century. That idea may at one time have fit the facts well, so the thinking goes, but now, with the explosion of biological knowledge attendant to the beginning of the twenty-first century, it clearly can be seen that this old idea will no longer work and it is time to discard it. The idea, of course, is evolution.

Behe's argument, as summarized in his 1996 book, *Darwin's Black Box*,[3] is that Darwinian evolution simply cannot account for the complexity of the living cell.

Charles Darwin worked in almost total ignorance of the fields we now call genetics, cell biology, molecular biology, and biochemistry, as Behe correctly notes. These disciplines, and the enormous amount of data they now generate, have come to dominate modern biology. They now give us a detailed picture of the living cell, a picture that shows the intricate complexity of life itself in a way that Darwin could not have anticipated. In light of so much new data, it is high time, according to Behe, to ask whether or not evolution is still an adequate explanation.

Surprisingly, he seems to think that evolution is what the residents of Lake Wobegone might call a "pretty good theory." In fact, he goes out of his way to make clear that he agrees with a great deal of what comes under the heading of evolution:

> For the record, I have no reason to doubt that the universe is the billions of years old that physicists say it is. Further, I find the idea of common descent (that all organism share a common ancestor) fairly convincing, and have no particular reason to doubt it. I greatly respect the work of my colleagues who study the development and behavior of organisms within an evolutionary framework, and I think that evolutionary biologists have contributed enormously to our understanding of the world.[4]

So, what does Behe actually disagree with? Although he acknowledges that nineteenth century Darwinism successfully explained the origin of many complex structures, the inner workings of the cell remained a mystery, a "black box," to evolutionary theorists for many years. But then, in the twentieth century, a revolution in biochemistry that made life approachable at the molecular level changed the playing field for evolution. As Behe tells it:

> Just as biology had to be reinterpreted after the complexity of microscopic life was discovered, neo-darwinism must be reconsidered in light of advances in biochemistry. The scientific disciplines that were part of the evolutionary synthesis are all nonmolecular. Yet for the Darwinian theory of evolution to be true, it has to account for the molecular structure of life. It is the purpose of this book to show that it does not.[5]

Wouldn't it be exciting if my own discipline, cell biology, could help another field of biology to break out of a centuries-old intellectual stranglehold? A nonscientist shouldn't underestimate how appealing this would be to a cell biologist like me, who believes—apparently along with Behe—that the keys to understanding all of life can ultimately be found at the cellular level.

Naturally, the first question for Behe has to be *why* the Darwinian theory of evolution cannot account for the molecular structure of life.

His answer is that the cell is loaded with systems that are "irreducibly complex," and that evolution cannot produce them.

> By irreducibly complex I mean a single system composed of several well-matched, interacting parts that contribute to the basic function, wherein the removal of any one of the parts causes the system to effectively cease functioning.[6]

If all the parts of a system are needed to make it work—and you can test that by simply taking one of them away—then a system is, according to Behe, irreducibly complex. By this definition, the cell seems to be loaded with irreducibly complex systems that do everything from copying DNA to responding to light by moving cells around. But why does this have anything to do with evolution? Here Behe proposes something truly radical—he claims that irreducibly complex systems *cannot* be produced by evolution.

Why not? Because the mechanism of evolution is driven by natural selection:

> Since natural selection can only choose systems that are already working, then if a biological system cannot be produced gradually it would have to arise as an integrated unit, in one fell swoop, for natural selection to have anything to act on.[7]

In other words, if a system requires *all* of its parts to have a useful function, then there is no way for natural selection to produce it. This requires a bit of explanation. Let's suppose, for example, that the cell contains a tiny microscopic machine that enables it to swim, and let's further suppose that the cell's swimming machine contains twenty different, carefully matched parts. Like much of the machinery of any cell, those parts are made of proteins, and each protein is specified by its own gene.

Where did the swimming machinery come from? The stock answer, as Behe notes, is that it must have been produced by evolution. Cells that could swim must have had a selective advantage over cells that could not, and therefore the genes that could produce the swimming machinery were favored by natural selection.

This seems like a fine explanation until, according to Behe, we begin to sweat the details. Suppose that the swimming machinery is irreducibly complex, meaning that if we remove or damage just one of those twenty proteins, it doesn't work. Now we have a real problem. Evolution, according to Darwin, produces change a little bit at a time, with natural selection honing the characteristics of organisms (and cells) to make them better and better. But a swimming machine with only sixteen or seventeen proteins, no matter how good those parts are, won't work. So a partially assembled swimming machine cannot be favored by natural selection. In fact, even a cell with nineteen of the twenty proteins in perfect working order has no selective advantage at all over a cell with none of them. That would seem to mean that the only way that evolution could have produced the swimming machinery was to assemble twenty very, very lucky macromutations together in the same cell at the same time so that the ability to swim arose instantly. Even evolutionists agree that the chances of that happening, of stringing together so many lucky mutations to produce instantaneous function in a complex structure, are just too remote to be taken seriously.

> An irreducibly complex system cannot be produced directly by numerous, successive, slight modifications of a precursor system, because any precursor to an irreducibly complex system that is missing a part is by definition nonfunctional.[8]

This means, according to Behe, that the theory of evolution is in serious trouble. Even more trouble than I have indicated. My twenty-protein swimming machine was hypothetical. A real swimming machine, such as the flagellum found in many bacterial cells, contains more than fifty parts. Cilia and flagella found in eukaryotic cells, cells that (like our own) contain nuclei, are even more complex, probably containing as many as 250 distinct proteins.

And that's just the beginning. As Behe correctly points out, the cell is filled with complex protein machines. These include the pathway of signaling proteins in the light-sensitive cells of our eyes, the intricate cascade of proteins that cause blood to clot, the incredibly flexible antibody-producing machinery of our immune systems, and the vital chemical pathways that are used to produce compounds essential for each living cell.

How could evolution ever have produced even one of these irreducibly complex machines, let alone all of them? According to Behe, it couldn't have. His argument, the great revelation of *Darwin's Black Box,* was that all that complexity must have come from somewhere else.

As appealing as all of this was, as I read Behe's book I began to get the impression that this seemingly new argument against natural selection had a familiar ring. Displaying a certain lack of modesty, Behe let the cat out of his black box:

> The result of these cumulative efforts to investigate the cell—to investigate life at the molecular level—is a loud, clear, piercing cry of "*design!*" The result is so unambiguous and so significant that it must be ranked as one of the greatest achievements in the history of science. The discovery rivals those of Newton and Einstein, Lavoisier and Schrödinger, Pasteur, and Darwin.[9]

Clearly, Michael Behe would like us to believe that he has discovered a new biological principle called "design." But the real news in *Darwin's Black Box* is not design. Nor is it that Newton and Einstein have been joined by a biochemist in the scientific pantheon. It is, instead, that a classic argument from the nineteenth century could be attractively rewrapped in the shiny packaging of biochemistry. At least for a while, many would fail to recognize just how old this argument really is.

Old Reliable

Although Michael Behe has endowed it with a new term—"irreducible complexity"—the heart and soul of his treatise against evolution is neither new nor novel. It is the "argument from design," the oldest, the most compelling, and quite simply the best rhetorical weapon against evolution.

What is the argument from design? Living things are complex and intricate structures. How complex? Study the literature of any aspect of biology—anatomy, physiology, genetics, even molecular biology—and you will find scientists working to understand the "design" of a particular structure, pathway, or mechanism. In the physical world, when we stumble across an intricate structure composed of multiple working parts—say, a pocket watch—we quickly assume that it was designed. Living things, as the argument goes, are even more complex. Therefore, it is clearly logical to conclude that they, too, were designed. Not evolved, but designed!

Behe openly admires the great nineteenth century advocate of design,

Rev. William Paley, and quotes extensively from Paley's *Natural Theology* to make his case. Historians of science will immediately recognize the problem here. *Natural Theology* was published in 1803, and Paley's version of the argument from design was well known to Darwin. He considered it carefully when he wrote *The Origin,* published more than a half century later. And Darwin answered the argument. To do this, he chose the eye—a complex, multipart organ that was itself favored as an example by advocates of the design argument:

> To suppose that the eye, with all its inimitable contrivances for adjusting the focus to different distances, for admitting different amounts of light, and for the correction of spherical and chromatic aberration, could have been formed by natural selection, seems, I freely confess, absurd in the highest possible degree.[10]

Since the eye could not work properly without all of its parts in place, including the lens, retina, iris, optic nerve, and many others, it is a classic example of an irreducibly complex organ. So, presumably it presents a terrible problem for evolution. But Darwin already had the answer:

> Yet reason tells me, that if numerous gradations from a perfect and complex eye to one very imperfect and simple, each grade being useful to its possessor, can be shown to exist; if further, the eye does vary ever so slightly, and the variations be inherited, which is certainly the case; and if any variation or modification in the organ be ever useful to an animal under changing conditions of life, then the difficulty of believing that a perfect and complex eye could be formed by natural selection, though insuperable by our imagination, can hardly be considered real.[11]

Darwin's reasoning cuts right to the heart of the argument from design. It boldly claims that the interlocking complexity of a multipart organ like the eye could indeed be produced by natural selection. How? As Darwin noted, all that we really need to show is existence of "numerous gradations" from the simple to the complex. Then all natural selection has to do is to favor each step in the pathway from simple to complex, and we have solved the problem.

The crux of the design theory is the idea that by themselves, the individual parts or structures of a complex organ are useless. The evolutionist says no, that's not true. Those individual parts can indeed be useful, and it's by working on those "imperfect and simple" structures that natural selection eventually produces complex organs. How does the argument end? Well, it hasn't completely ended (which is one of the reasons I have written this chapter!), but within the scientific community Darwin has won.

In the case of the eye, biologists have realized that any ability, no matter how slight, to sense light would have had adaptive value. Bacteria and algae, after all, manage to swim to and from the light with nothing more than an eyespot—a lensless, nerveless cluster of pigments and proteins. Zoologists have discovered numerous gradations of light-sensing systems in nature, most of which were far simpler and less perfect optically than the vertebrate eye. The existence of so many working "pseudo-eyes" and "semi-eyes" in nature convinced natural scientists that Darwin's imagined intermediates between primitive light-sensing systems and complex eyes were feasible and real. The human eye, with all of its marvelous complexity, could indeed have been formed by evolution, one step at a time, from a series of simpler but still functioning organs, shaped every step of the way by natural selection.

Despite Darwin's success, in this century the argument from design has been trotted out again and again, in one form after another, as though it presents in each case a brand-new challenge for evolution. It doesn't, of course, but the persistence of evolution's critics requires them to appear to have discovered new and up-to-date scientific data against evolution. And no single argument in their crusades remotely approaches the cogent appeal of old reliable, the argument from design.

What Michael Behe has done is very much in this tradition. He literally has dusted off the argument from design, spiffed it up with the terminology of modern biochemistry, and then applied it to the proteins and macromolecular machines that run the living cell. Once we've figured this out, we can ask the key question: Is there anything really so different about proteins and cells that makes the argument from design work better at their level than it does at the level of the organism? We'll address this question below, but as the reader may suspect, the argument from design is about to take another fall.

The Watchmaker at Work

Paley's classic formulation of the argument from design compared the designed complexity of a manufactured watch to the far greater undesigned complexity of the living cell. Although Darwin answered this argument directly in *The Origin,* it persists so often in modern forms that in 1986 Richard Dawkins, a prominent evolutionary biologist from Oxford University, devoted a whole book to answering it yet again.

Taking Paley's challenge head-on, he called his book *The Blind Watchmaker,* a title that reflected the common metaphor of evolution as an agent that produces complex, even elegant design without direct intent or plan. To make his point, Dawkins chose another example of an intricate, complex, well-designed series of structures—the echolocation system of bats.

Bats have sonar, a system of sound-based detection "invented" by bats tens of millions of years before engineers thought of similar systems to fix the location of underwater objects like submarines. The sophistication of bat sonar systems is dazzling, involving the complex interlocking of specialized sound-making and sound-sensing systems. How could blind, random, undirected evolution possibly have produced such an intricate set of structures and organs, so brilliantly dedicated to a single purpose? That was exactly the challenge that Dawkins took up.

According to Behe's criteria, the echolocation system is irreducibly complex and therefore could not be produced by evolution. After all, what good would it do a bat to evolve the ability to *produce* its high-pitched sonar pulses unless it also had the ability to *detect* them?

As Dawkins explained, following Darwin's lead, the different parts of any complex system, including an echolocation apparatus, evolve *together* in a series of working stages. We begin with ordinary sound-making and hearing organs, gradually improving, one step at a time, until they reach the bat's current level of complex engineering.

The design theorist would argue that to support that explanation you would have to show two things: first, that a crude sort of echolocation system could be shown to work—at least a little—without any special adaptations at all; and second, that each stepwise improvement in a series of intermediate stages from the simple to the complex would actually produce functioning systems that could be favored by natural selection.

Doing that is easier than you might think. Even animals with famously unspecialized auditory systems—like ours—have a limited ability to echolocate. A crude, working version of the complex function is already present in us, and in most other animals. This gives natural selection something to start with, and under the right conditions—a habitat and lifestyle in which any improvement in echolocation would be advantageous—the evolution of an improved, more complex system is only a matter of time. This hasn't happened in humans for the simple reason that we neither fly nor swim in darkness. In our normal habitats stereoscopic vision works very well, but surely this is not true of all animals. If the account I have just given is even remotely true, then we should expect that many animals, not just bats, should have evolved echolocation systems. And so they have, many times, in fact. By any standard, as Dawkins points out in *The Blind Watchmaker,* the evolution of the bat's complex sonar system can be fully explained by natural selection.

It would be nice if we had an actual physical record of the evolution of these systems, but the process of fossilization doesn't always cooperate. This is certainly so in the case of bats, which are rarely preserved in fossils, even from geologically recent times. This does not mean that we don't have well-documented records of the evolution of other complex systems, and the auditory apparatus itself is a perfect case in point.

The three smallest bones in the human body—the *malleus*, *incus*, and *stapes*—carry sound vibrations across the middle ear, from the membrane-like tympanum (the eardrum) to the oval window. This five-component system perfectly fits the criterion of irreducible complexity—if any one of its parts is taken away or modified, hearing is lost. This is the kind of system that evolution supposedly cannot produce, since, as Behe has said, it would have to "arise as an integrated unit, in one fell swoop." Unfortunately for design theorists, the fossil record elegantly and precisely documents exactly how this system formed.

During the evolution of mammals, over several million years, two of the bones that originally formed the rear portion of the reptilian lower jaw were gradually pushed backwards and reduced in size until they migrated into the middle ear, forming the bony connections that carry vibrations into the inner ears of present-day mammals. This is an example of a system of perfectly formed interlocking components, specified by multiple genes, that was gradually refashioned and adapted for another purpose altogether—something that evolution's critics claim to be impossible. Evidence of this sequence, which was worked out in detail by Arthur W. Crompton of Harvard University, even includes a fossil species possessing a remarkable double articulation of the jaw joint—an adaptation that allowed the animal both to eat and hear during the transition, enabling natural selection to favor each of the intermediate stages.[12]

Arriving at the irreducible complexity of the five-part auditory apparatus was easy for evolution because it doesn't work the way that its critics claim it must. To fashion the three-bone linkage that conducts sound from the eardrum, evolution didn't have to start with a nonworking incomplete one-bone or two-bone middle ear. Instead, it started with a perfectly good working reptile-style ear, which had a single internal bone. Then it grabbed two other bones from a different organ, the jaw, and used them to expand and improve the apparatus.

Remember Behe's statement that "any precursor to an irreducibly complex system that is missing a part is by definition nonfunctional"? Well, there's just no other word for it—that statement is *wrong*. What evolution does is to add parts that expand, improve, and sometimes completely refashion living systems. Once the expansion and remodeling is complete, every part of the final working system may indeed be necessary, just like the *malleus*, *incus*, and *stapes*. That interlocking necessity does not mean that the system could not have evolved from a simpler version—and in this case we know that is exactly what happened.

This, in essence, is how the watchmaker works, beginning with a set of basic structures and gradually modifying them to produce new functions from old parts. Dawkins's exceptionally readable accounts of how evolution produces *apparent* design is an excellent answer to those who claim certainty that Darwinism cannot account for the organs of extreme perfection that are common to living things. In fact, evolution does exactly that.

But one question remains. Is there something special about biochemistry that prevents evolution from doing exactly the same thing to a microscopic system composed of proteins? As Michael Behe has said:

> Until recently, molecular biologists could be unconcerned about the molecular details of life because so little was known about them. Now the black box of the cell has been opened, and the infinitesmal world that stands revealed must be explained.[13]

I agree; in fact, I agree enthusiastically. So let's take a look.

Opening the Box

The challenge that Behe offers to evolution is straightforward, and it comes from an honorable source—Charles Darwin, the man himself. In Chapter 6 of *The Origin,* "Difficulties of the Theory," he wrote:

> If it could be demonstrated that any complex organ existed, which could not possibly have been formed by numerous, successive, slight modifications, my theory would absolutely break down.[14]

Behe argues that the cell is filled with biochemical machines that are irreducibly complex, and so could not have been formed by those numerous, successive, slight modifications that Darwin required. Therefore, in Behe's view, the theory has indeed broken down. The logical chain of his argument begins with the claim of irreducible complexity, and from it follows the conclusion of design. Here's one example of how he applies that reasoning to a complex, cellular structure like the cilium:

> But since the complexity of the cilium is irreducible, then it can not have functional precursors. Since the irreducibly complex cilium can not have functional precursors it can not be produced by natural selection, which requires a continuum of function to work. Natural selection is powerless when there is no function to select. We can go further and say that, if the cilium can not be produced by natural selection, then the cilium was designed.[15]

There are many ways to answer Behe's arguments, and most of these have already found their way into print. As a cell biologist, I was particularly amused by a biochemist's suggestion that the complexity of the cilium is irreducible. A cross-section of the kind of cilium Behe has in mind does reveal a structure of exceptional beauty and complexity (see Figure 5.1). Nine pairs of microtubules surround two central microtubules, each joined by an intricate series of spokes and linkers. Because textbooks say that the "9+2" structure is found in everything from single-celled algae to human sperm, a biochemist might easily have assumed that this particular pattern was the only one that worked, hence the conclusion of irreducible complexity.

A phone call to any biologist who had ever actually studied cilia and flagella would have told Behe that he's wrong in his contention that the 9+2 structure is the *only* way to make a working cilium or flagellum. Comparative studies on a wide variety of organisms (as in Figure 5.2) show that there are many ways to make a working cilium or flagellum without some of the parts that Behe seems to believe are essential.

Sperm from the caddis fly *Polycentropus* are different—they have a 9+7 arrangement in which the central pair of microtubules, which Behe believes is essential, is replaced by a cluster of seven microtubules. Mosquitoes of the genus *Culex* don't have a central pair at all—they have just a single microtubule in the center, making them 9+9+1. Eel sperm

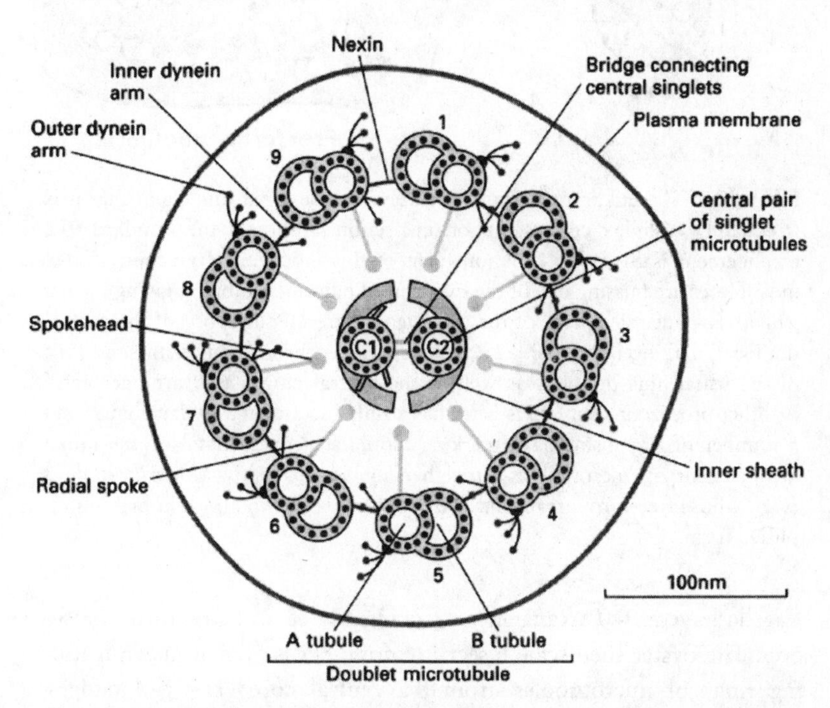

Figure 5.1. Detailed cross-section of a eukaryotic cilium. Cilia are whip-like structures that cells use to generate force and movement. Microtubules, arranged in a characteristic 9+2 pattern, form the core of most eukaryotic cilia. This diagram, from a leading textbook on cell biology, emphasizes the structural complexity of the 9+2 arrangement.

Figure 23–29 (a) from H. Lodish *et al., Molecular Cell Biology*, © 1995, Scientific American Books, New York. Used by permission.

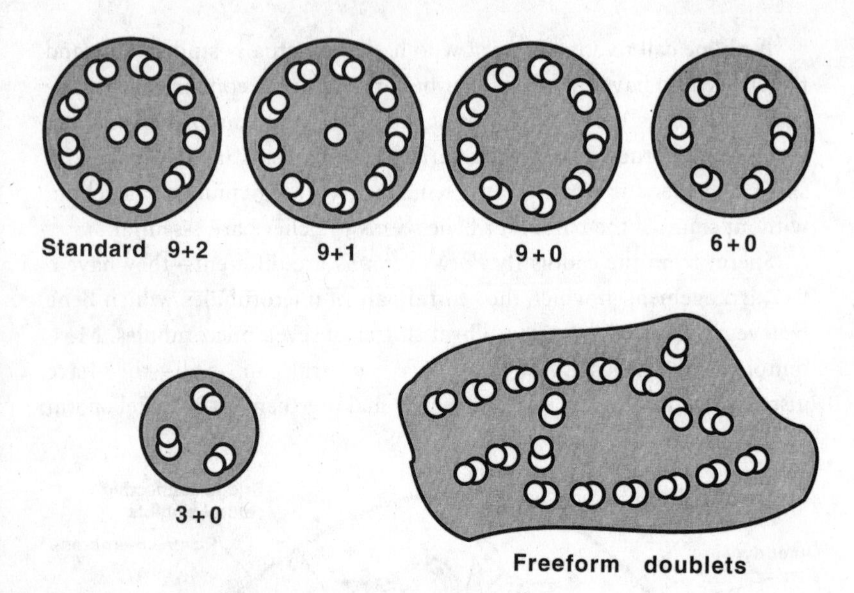

Figure 5.2. The eukaryotic cilium is surely complex, but any claim that it is irreducibly complex collapses upon inspection. Although the standard 9+2 arrangement is the most common form of this biochemical machine, *Culex* mosquitoes are missing one of the two central pair microtubules, giving them a 9+1 arrangement.[16] Sperm of the eel *Anguilla* are missing *both* of the central doublets, making them 9+0.[17] *Lecudina tuzetae,* a protozoan, is missing three of the usual nine doublets as well as the central pair, a 6+0 arrangement.[18] Another protozoan, *Diplauxis hatti,* has a fully functional 3+0 flagellum,[19] and a number of organisms have working motile structures that lack the radial arrangement of microtubules altogether, including the fly *Monarthropalpus buxi*, whose free-form arrangement of doublets is shown above in highly simplified form.

flagella have a 9+0 arrangement, lacking any central structure whatsoever. The oyster shell scale insect *Lipidosaphes* is even stranger, featuring rings of microtubules around a central core. The gall midges, including *Clinodiplosis,* have a bizarre free-form arrangement in which dozens of microtubule doublets are scattered throughout the sperm tail. And finally, the protozoan *Diplauxis hatti* has a 3+0 arrangement in which many of the supposedly indispensable parts of the 9+2 structure are missing.[20]

In nature, we can find scores of cilia lacking one or more of the components supposedly essential to the function of the apparatus. It shouldn't be surprising that these novel structures do not all work exactly like the 9+2 arrangement. Nonetheless, they all do work, and each one is successful in its own way. These are sperm cells, after all, and the price for a gene that produces a defect in sperm function is instantaneous elimination from the next generation. What we actually see among cilia and flagella in nature is something entirely consistent with Darwin's call for numerous gradations from the simple to the complex. Once we have found a series of less complex, less intricate, differently organized flagella, the contention that this is an irreducibly complex structure has been successfully refuted.

I suppose we could close the case right there. After all, the existence of so many simpler molecular machines in protozoans and eels and gall midges shows that Behe's central thesis, that the flagellum could not have "functional precursors," is disproved, and that's that. But let's look a little further.

EVOLVING COMPLEXITY

I want to continue to examine biochemical evolution because Behe's central contention is one that I enthusiastically endorse. If Darwinism cannot explain the interlocking complexity of biochemistry, then it is doomed. So the real test is a simple one—can Darwinian evolution explain the development of complex biochemical machines? Let's find out.

We have already seen in Chapter 4 that the evolution of new proteins can be observed in the laboratory. An interesting aspect of this phenomenon is how closely we can watch it happening. In 1997, a group of scientists in California decided to watch the evolution of a new interface between two proteins.[21] They did this with an important signaling molecule, human growth hormone. This molecule, itself a protein, binds to cells by means of receptors, precisely shaped molecules on cell surfaces that fit the three-dimensional shape of the hormone. The fit between hormone and receptor is almost like the fit between a key and a lock—break off even a tiny part of the key, and it may no longer turn the lock.

To study the fit between the hormone key and the receptor lock, these researchers figured they'd break off part of the receptor. They deleted a tryptophan, a large amino acid found at the exact point in the receptor where the hormone binds. Having lost that amino acid, the receptor's shape changed and it could no longer bind the hormone. The key would no longer fit the lock. Now the researchers let evolution take over.

Using the latest genetic technology, they randomly mutated the coding regions for five amino acids in the growth hormone, generating roughly 10 million different mutant combinations. These mutant growth hormones were then selected to find ones that could bind to the previously mutated receptor. Was this too much to ask of a random process? Would we be expecting too much for a single mutation or even a series of mutations to refashion the hormone so that it now fit the mutated receptor? Was evolution stuck? Not at all. Their random mutation strategy generated a new version of the hormone that fit the mutated receptor nearly one hundred times tighter than the nonmutant version. Those random mutations so scorned by the critics of evolution are the source of what produces what looks for all the world like a perfectly *designed* fit between hormone and receptor.

Watching the shapes of the protein as it changed did the usual study on protein evolution one better. To the investigators' very great surprise, the new parts of the hormone emerged from portions of the protein they admitted they never would have thought of as a source of mutations to restore the fit between hormone and receptor. The investigators wrote that these experiments "reveal how large, functional changes can be rescued by a limited number of mutations." More to the point, it illustrates how two proteins, two parts of a biochemical machine, can evolve together.

A study like this shows that evolution can act in unexpected, unanticipated ways to fashion novel proteins. And it shows—in remarkable detail—how protein-to-protein interactions are maintained. So long as selective pressure, even slight selective pressure, exists, it will hone and refine the interactions between two proteins of any biochemical machine.

Yes, I can see the skeptic nodding, admitting that this work shows how one protein can be remodeled to fit another. But how do we approach the larger question? How do we evolve a system of multiple parts? Good question. Let's take a look at that next.

Parts Is Parts

What happens if we step things up to the next level, asking evolutionary mechanisms to develop a multipart system, a system that, in Behe's terminology, would be irreducibly complex? The most direct way to do this would be with a true acid test—by using the tools of molecular genetics to wipe out an existing multipart system and then see if evolution can come to the rescue with a system to replace it.

The "lac" genes of bacteria are just such a system. Lactose is a sugar, similar in many ways to sucrose, ordinary table sugar, that bacteria can use as a food. In order to do this, they first have to produce an enzyme that can cut lactose in half, releasing two simple sugars, glucose and galactose, that the metabolism of the cell can use for energy. Not surprisingly, bacteria are "smart" enough to "know" that there is no point in making this enzyme—known as galactosidase—when there is no lactose in the medium in which they are growing. In the absence of lactose they automatically shut off the genes required to utilize lactose for food. Clever biochemistry. Behe might say, "Clever design."

The bacterium pulls this feat off by combining highly sensitive control genes with the structural genes that actually specify the amino acids in the galactosidase enzyme. The control gene keeps the enzyme gene switched off except when it needs to produce enough of the enzyme to metabolize lactose. Could evolution have produced such a lovely two-part system? Barry Hall[22] tested this possibility in 1982 by deleting the structural gene for galactosidase. He then "challenged" cells with the deletion to grow on lactose. At first, of course, they couldn't. Before long, however, mutant strains appeared that could handle lactose nearly as well as the originals.

How could this be? How could these cells have reconstructed the information from the missing gene in such a short time, using only the random, undirected processes of mutation and natural selection? The answer is that these bacteria didn't make the new galactosidase enzyme from scratch. They made it by tinkering with another gene, in which a simple mutation changed an existing enzyme just enough to make it also capable of cleaving the bond that holds the two parts of lactose together. Now, you might think that this wouldn't be enough, and you'd be right. Simply reengineering an existing protein to replace galactosidase would make no difference unless the control region of *that* gene was also changed to ensure that the gene was expressed when lactose was present. Significantly, when Hall looked at the control regions of the mutant replacement gene, he found that they had been mutated as well—some of them were switched on all the time, but a few of them responded directly to lactose, switching the gene on and off as needed.

That would have been impressive enough, but Hall's clever germs didn't stop there. When he selected them further to grow on another sugar, lactulose, he obtained a second series of mutants with a new enzyme that accidentally (in a sense) produced allolactose, the very same chemical signal that is normally used to switch on all of the lac genes. This important development meant that now the cells could switch on synthesis of a cell membrane protein, the lac permease, that speeds the entry of lactose into the cell. Summarizing this work, evolutionary biologist Douglas Futuyma wrote:

> Thus an entire system of lactose utilization had evolved, consisting of changes in enzyme structure enabling hydrolysis of the substrate; alteration of a regulatory gene so that the enzyme can be synthesized in response to the substrate; and the evolution of an enzyme reaction that induces the permease needed for the entry of the substrate. One could not wish for a better demonstration of the neoDarwinian principle that mutation and natural selection in concert are the source of complex adaptations.[23]

Think for a moment—if we were to happen upon the interlocking biochemical complexity of the reevolved lactose system, wouldn't we be impressed with the intelligence of its design? Lactose triggers a regulatory sequence that switches on the synthesis of an enzyme that then metabolizes lactose itself. The products of that successful lactose metabolism then activate the gene for the lac permease, which ensures a steady supply of lactose entering the cell. Irreducible complexity. What good would the permease be without the galactosidase? What use would either of them be without regulatory genes to switch them on? And what good would lactose-responding regulatory genes be without lactose-specific enzymes? No good, of course. By the very same logic applied by Michael Behe to other systems, therefore, we could conclude that the system had been designed. Except we *know* that it was *not* designed. We know it evolved because we watched it happen right in the laboratory!

No doubt about it—the evolution of biochemical systems, even complex multipart ones, is explicable in terms of evolution. Behe is wrong.

The Sound of Silence

One of the more striking charges in *Darwin's Black Box* is that the kind of study I have just described, a test of the ability of ordinary evolutionary processes to produce complex biochemical systems, has *never* been done. Obviously, Behe has overlooked a couple of papers. This is easy enough to understand. The literature of science is so large that it would be foolish of me to criticize Behe (or anyone else) for missing a paper here or there. But the charges he makes are stunning in scope and damning in character:

> There is no publication in the scientific literature—in prestigious journals, specialty journals, or books—that describes how molecular evolution of any real, complex biochemical system either did occur or even might have occurred.[24]

Behe claims to have scanned all of the issues of the well-regarded *Journal of Molecular Evolution,* but found nothing. A stunning silence in the face of what he regards as the "loud, piercing cry of design" at the level of the cell itself. The silence, he implies, is damning. Evolution is silent, even in its own journals, because it cannot answer the challenge of design.

Some of this silence, no matter what Behe says, makes good sense. Before you can explain a complex system, you've got to know what it's made of and how it works. Behe delights in talking about the bacterial flagellum, a marvelous biochemical machine that actually produces rotary motion from an electrochemical gradient across the cell membrane. Why haven't evolutionary biologists explained how this structure was produced? Well, one good reason is that no one yet fully understands how all its parts work. In 1998, fully three years *after* the publication of Behe's book, structural biologist David DeRosier reviewed our current understanding of the workings of the bacterial flagellum in the journal *Cell.* He concluded his state-of-the-art review by writing:

> The mechanism of the flagellar motor remains a mystery. It is not clear if there are any similarities of its mechanism to those used in eukaryotic cells. The differences are apparent but we may not know about similarities

> until we have more detailed structures for the motor parts; we look forward to the day when we do.[25]

Before evolution is excoriated for failing to explain the evolution of the flagellum, I'd request that the scientific community at least be allowed to figure out how its various parts work. Speculating on the evolution of a biochemical structure only partly understood is not going to lead to good science.

Nonetheless, if we really *do* understand how a biochemical machine is put together, I agree with Behe that there's no reason for evolution not to come up with a good explanation. And in quite a few cases, that's exactly what has happened.

I decided to take up Behe's challenge by having a quick look at the scientific literature published immediately after his book's appearance. If he was right, I should not have been able to find a single published Darwinian explanation for the origin of a biochemical machine. In fact, even a brief glance at the recent literature of molecular evolution turned up four glittering examples of what Behe claimed would never be found.

In January 1998, in the lead article in *American Scientist,* a journal of general science for technical audiences, Anthony M. Dean, described in great detail the structural and biochemical changes that took place millions of years ago to produce two enzymes, two biochemical machines, essential to the subsequent evolution of life. Both enzymes are known as isocitrate dehydrogenases (ICDH). To work its chemical magic, an ICDH needs another compound to help it out, a compound known as a cofactor, and there's the rub. There are only two known cofactors (NAD and NADP) and all known ICDHs use either one or the other. How could these two different forms of ICDH have evolved from just one ancestral form?

Dean has the answer. Using a theoretical framework called the "Neutral Theory" of molecular evolution (fashioned by Japanese scientist Mootoo Kimura), Dean explored how random changes in a few key amino acids could have effected dramatic shifts in enzyme activity. By further testing the structural adaptations caused by each of these mutations, Dean shows how natural selection could have supported each of the changes necessary to produce one enzyme from the other. Because ICDHs are essential to the metabolism of even the smallest cells, these adaptations were essential events in the early history of life, making it possible to test his ideas against the known structures of ICDHs found in scores of organisms, a test that both his analysis and evolution pass. Score one for the evolution of a true biochemical machine, an important enzyme.[26]

In 1997, John M. Logsdon and Ford Doolittle reviewed in detail how these same mechanisms could have produced, in strictly Darwinian fashion, the remarkable "antifreeze" proteins of Antarctic fish.[27] The novelty of this study is that it contained examples of how evolution could recruit introns, the noncoding regions found in the middle of many genes, to produce dramatic changes in the characteristics of proteins. Another point for evolution, this one a step up in subtlety and complexity.

Now for the evolution of a complex multipart biochemical structure. In 1998, Siegfried Musser and Sunney Chan[28] described the evolution of the cytochrome c oxidase protein pump, a complex, multipart molecular machine that plays a key role in energy transformation by the cell. In human cells, the pump consists of six proteins, each of which is necessary for the pump to function properly. It would seem another perfect example of irreducible complexity. Take one part away from the pump, and it no longer works. Yet these authors were able to produce, in impressive detail, "an evolutionary tree constructed using the notion that respiratory complexity and efficiency progressively increased throughout the evolutionary process."

How is this possible? If you believed Michael Behe's assertion that biochemical machines were irreducibly complex, you might never have bothered to check; and this is the real scientific danger of his ideas. Musser and Chan did check, and found that two of the six proteins in the proton pump were quite similar to a bacteria enzyme known as the cytochrome bo3 complex. Could this mean that *part* of the proton pump evolved from a working cytochrome bo3 complex? Certainly.

Critics of evolution have always loved to ask, "Of what use is half a wing?" The implication, of course, is that evolution couldn't possibly fashion complex organs like wings one step at a time because the wing acquires its function as a flying organ only when it is complete. That's not true. It's easy to show living examples of forelimbs only partly modified for flight that have useful functions as gliding appendages. Half a wing, under the right circumstances, can be very useful. To evolution, the beauty of biochemistry is that it's even easier to demonstrate the value of half a protein machine.

An ancestral two-part cytochrome bo_3 complex would have been fully functional, albeit in a different context, but that context would indeed have allowed natural selection to favor its evolution. How can we be sure that this "half" of the pump would be any good? By referring to modern organisms that have full, working versions of the cytochrome bo_3 complex. Can we make the same argument for the rest of the pump? Well, it turns out that each of the pump's major parts is closely related to working protein complexes found in microorganisms. This shows, once again, that the notion of irreducible complexity is nonsense. Evolution assembles complex biochemical machines, as Musser and Chan proposed, from smaller working assemblies that are adapted to fit novel functions. The multiple parts of complex biochemical machines are themselves assembled from smaller, working machines developed by natural selection.

Last of all, remember Behe's assertion regarding the *Journal of Molecular Evolution*. If any journal dealt with the evolution of biochemical systems, this should be the one, but, he writes, "In fact, none of the papers in *The Journal of Molecular Evolution* over the entire course of its life as a journal has ever proposed a detailed model by which a complex biochemical system might have been produced in a gradual, step-by-step Darwinian fashion."[29] If that statement was correct when *Darwin's Black Box* was published early in 1996, by the end of that year it was clearly in error.

In 1996, Enrique Meléndez-Hevia and his colleagues published just such a paper in the *Journal of Molecular Evolution*. As every high school biology student learns, the Krebs cycle is an extremely complex series of interlocking reactions that release chemical energy from food. If there was any doubt that this paper represented exactly the kind of study that Behe demanded, the authors removed it in their first two sentences:

> The evolutionary origin of the Krebs citric acid cycle has been for a long time a model case in the understanding of the origin and evolution of metabolic pathways. How can the emergence of such a complex pathway be explained?[30]

This paper is a head-on refutation of Behe's assertion that evolution cannot account for the development of complex biochemical systems. Its authors applied the tests that Behe demands to their evolutionary schemes—the requirement that each and every intermediate stage be functional and favored by natural selection. Were they successful? Absolutely. And their success was due in no small measure to their realization that the parts of the Krebs cycle could be selected separately, adapted first to different biochemical functions that had nothing to do with the eventual chemistry of the complete cycle. In other words, just like the evolution of the cytochrome c oxidase protein pump, the individual parts of the complex machine appear first as functional units, which then are borrowed, loaned, or stolen for other purposes:

> The Krebs cycle has been frequently quoted as a key problem in the evolution of living cells, hard to explain by Darwin's natural selection: How could natural selection explain the building of a complicated structure in toto, when the intermediate stages have no obvious fitness functionality? This looks, in principle, similar to the eye problem, as in "What is the use of half an eye?" (see Dawkins 1986, 1994). However, our analysis demonstrates that this case is quite different. The eye evolved because the intermediary stages were also functional as eyes, and, thus the same target of fitness was operating during the complete evolution. In the Krebs cycle problem the intermediary stages were also useful, but for different purposes, and, therefore, its complete design was a very clear case of opportunism. The building of the eye was really a creative process in order to make a new thing specifically, but the Krebs cycle was built through the process that Jacob (1977) called "evolution by molecular tinkering," stating that evolution does not produce novelties from scratch: It works on what already exists. The most novel result of our analysis is seeing how, with minimal new material, evolution created the most important pathway of metabolism, achieving the best chemically possible design. In this case, a chemical engineer who was looking for the best design of the process could not have found a better design than the cycle which works in living cells.[31]

How does evolution produce a biochemical pathway that *looks* as though it were designed? The answer is right here, and it's the same answer we've seen in other systems as well—by modifying existing structures and proteins to produce new organs, new functions, and especially, new biochemistry.

The Krebs cycle is a complex biochemical pathway that requires the interlocking, coordinated presence of at least nine enzymes and three cofactors. And a Darwinian explanation for its origin has now been crafted. Against the backdrop of these and many other studies, the claim that Darwinism has failed to explain "even a single biochemical machine" sounds a little hollow.

Plugging a Leak

By now it should be clear that any claim that evolution cannot produce complex, well-designed biochemical machines is just plain wrong. We could, I think, fairly and reasonably close the books on the issue of irreducible complexity, and move on to topics of genuine scientific interest. In the interest of giving the insurgent point of view every benefit of the doubt, however, let's go one more round—but this time we'll do it on the critic's home turf.

One of Behe's favorite examples against evolution, for him a true case of an irreducibly complex biochemical machine, is the process of blood clotting, or coagulation as it is sometimes called. In *Darwin's Black Box,* Behe details the intricate series of factors, activators, and chemical reactions used to start and control the process of blood clotting. This collection of proteins, more than a dozen in all, circulate in plasma, the fluid portion of the blood, just waiting for an appropriate stimulus to set them off. As he describes it, the clotting process is a "Rube Goldberg machine" of bewildering complexity that requires each and every step to be performed perfectly to be effective. And it could not have evolved, according to Behe, because the whole system, each and every part, has to be present in order for the system to work.

This is a remarkable assertion, especially given the fact that Behe describes, almost admiringly, the years of extraordinary work done by molecular biologist Russell Doolittle on the evolution of the clotting mechanism. Despite this work, Behe claims that "no one on earth has the vaguest idea how the coagulation cascade came to be."[32]

Needless to say, all this came as a bit of a surprise to Doolittle himself. After reading Behe's book, he wrote about it in the *Boston Review,* noting that his early interests in the evolution of blood clotting had led him to devote thirty-five years of research to the general subject of proteins and the evolution. He wrote (with just a twinge of sarcasm), "Now it appears that I have wasted my career."

Well, Russell Doolittle, a distinguished member of the National Academy of Sciences, hasn't wasted his career. In fact, the human clotting system lends itself to exactly the kind of evolutionary analysis that seems impossible to Michael Behe.

The importance of the clotting process ought to be self-evident. Every time we cut ourselves, this system comes into play to prevent excessive loss of blood. Clotting saves us every time a bump or bruise causes blood vessels to break within the body. Who needs to have a clotting system? Well, we do, but so does any animal with a closed circulatory system, one in which blood flows through vessels. As it turns out, other mammals and most vertebrates have clotting systems that are nearly identical to ours.

At its core, the actual mechanism of clotting is remarkably simple. A fibrous, soluble protein called fibrinogen ("clot-maker") constitutes about three percent of the protein in blood plasma. Fibrinogen has a sticky portion near the center of the molecule, but the sticky region is covered by little amino acid chains with negative charges. Because like charges repel, these chains keep fibrinogen molecules apart.

When a clot forms, a protease (protein-cutting) enzyme called thrombin clips off the charged chains. This exposes the sticky parts of the molecule, and suddenly fibrinogens (which are now called fibrins) start to stick together, beginning the formation of a clot. To start the reaction, thrombin itself must be activated by another protease called Factor X. Believe it or not, Factor X requires two more proteases, Factor VII and Factor IX, to switch it on; and still other factors are required to activate *them* (see Figure 5.3). Why so many steps? The multiple steps of the clotting cascade *amplify* the signal from that first stimulus. If a single active molecule of one factor could activate, say, twenty or thirty molecules of the next factor, then each level of the cascade would multiply the effects of a starting signal. Put five or six steps in the cascade, and you've amplified a biochemical signal more than a million times. Clotting with fewer steps would still work, but it would take longer to produce a substantial clot, and would be much less responsive to smaller injuries.

Michael Behe properly emphasizes the intricacy of this process. He is correct to point out that if we take away part of this system, we're in trouble. Hemophiliacs, for example, are unable to synthesize the active form of an essential step in the cascade known as Factor VIII. This makes them

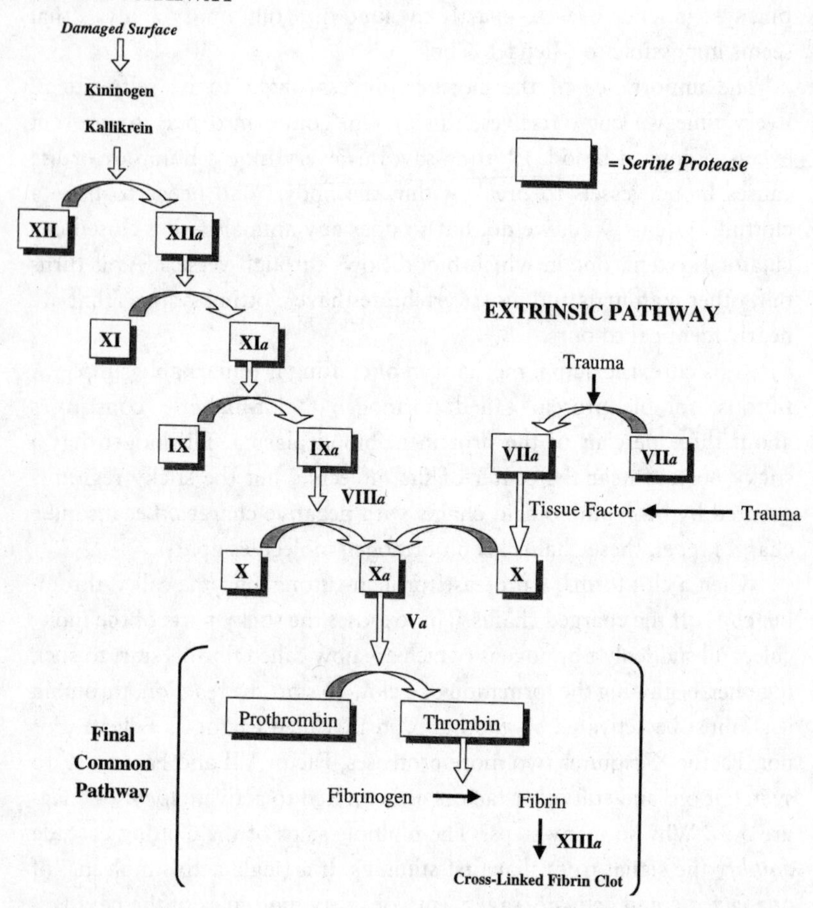

Figure 5.3. Blood clotting cascade. The formation of a fibrin clot can be triggered either by an "extrinsic" pathway, so named because factors released by tissue trauma are added to the blood, or by an "intrinsic" pathway, which does not require outside factors. Serine proteases, produced by gene duplication, form the core of both pathways and are highlighted in the diagram.

Redrawn from L. Stryer's *Biochemistry* (4th ed.), (New York: W.H. Freeman, ©1995), p. 253.

unable to complete the final step of one of the clotting pathways, and that's why hemophilia is sometimes known as the "bleeder's disease."[33] Defects or deficiencies in any of the other factors are equally serious. No doubt about it—clotting is an essential function and it's not something to be messed with. But does this also mean that it could not have evolved? Not at all.

The striking thing about this particular Rube Goldberg machine is how similar most of its parts are. Nearly all of the regulatory molecules belong to a single class of protein-cutting enzymes known as "serine proteases," and that, as Russell Doolittle realized many years ago, is the clue to understanding the system's evolution beginning with organisms that lacked a protein-based clotting system.

So far as we know, protein-based clotting systems are found only in vertebrates, crustaceans (animals such as crabs and lobsters), and other arthropods (including the horseshoe crab). Does this mean that other animals, such as starfish, sea urchins, and worms, automatically bleed to death every time one of their blood vessels is cut? No, and for two good reasons. First, their circulatory systems are under relatively low pressure, and this helps to minimize blood loss. Second, and more significant, all of these organisms have one form or another of "sticky" white cells in their bloodstreams. This means, just as the name implies, that some of their blood cells are able to stick to certain materials and form clumps. If a blood vessel is broken, white blood cells swept into the broken tissue quickly stick to tissue proteins like collagen. Over a few minutes' time, enough cells will stick to the exposed tissue proteins to plug the leak, and the integrity of the circulatory system is preserved.

To be sure, this isn't nearly as good a system as we have. It works slowly, and it's effective only when blood pressures are relatively low, as they are in most invertebrates. In fact, it's exactly the kind of "imperfect and simple" system that Darwin regarded as a starting point for evolution.

Could we object that even this simple system is too complex to have been produced by evolution? Not a chance. Those white cells are found in circulatory systems, open and closed, for a variety of purposes that have nothing to do with clotting. Once there, what would happen to them when the system sprang a leak? In a flash, these cells would pass from one kind of cellular environment, a fluid and dynamic one bounded by smooth walls and a liquid matrix, into a completely different one, an inflexible tissue environment packed with solid structures, fibrous extracellular proteins, and cable-like supporting molecules. Any change in the surface proteins and carbohydrates of the white cells that made them stick, even just a little bit, to that foreign matrix of tissue proteins would be favored by natural selection because it would help to seal leaks. At the molecular level, "sticking" is a matter of shape and charge. Therefore, any mutation that changed a neutral amino acid in a surface protein to a charged one, making the cell stick more tightly to a tissue protein like collagen, would be favored by natural selection.

This would produce a powerful selective pressure in favor of surface proteins that make white cells sticky. And given such pressure, it's hardly surprising that sticky white cells are widespread among animals—in fact, *all* animals with hearts, those muscular pumps that send blood coursing through the vessels of the circulatory system, have sticky white cells. Blood clotting is not an all-or-none phenomenon. Like any complex system, it can begin to evolve, imperfect and simple, from the basic materials of blood and tissue.

Blood plasma is already laden with soluble proteins for reasons like osmotic pressure[34] that have nothing to do with clotting. It turns out that serine proteases are abundant in the cells and tissues of the body, where they are used to transform and process biochemical signals—once again for reasons that have nothing to do with clotting. Therefore, when a broken blood vessel allows protein-rich plasma to flow into an unfamiliar environment, tissue proteases—quite accidentally—are now exposed to a new range of proteins, and they cut many of them to pieces. The solubility of these new fragments varies. Some are more soluble than the plasma proteins from which they were trimmed, but many are much less soluble. The result is that clumps of newly insoluble protein fragments begin to accumulate at the tissue-plasma interface, helping to seal the break and forming a very primitive clot. A simple, nonspecific clotting mechanism like this is used today by many invertebrates, so we know that it would have worked for the ancestor's of today's vertebrates.

What happened next? Here's where Doolittle's exceptional work has taken over. A series of ordinary gene duplications, many millions of years ago, copied some of these serine proteases. One of these duplicate genes was then mistargeted[35] to the bloodstream, where its protein product would have remained inactive until exposed to an activating tissue protease—which would happen only when a blood vessel was broken. From that point on, each and every refinement of this mechanism would be favored by natural selection. Where does the many-layered complexity of the system come from? Again, the answer is gene duplication. Once an extra copy of one of the clotting protease genes becomes available, natural selection will favor slight changes in the copy that might make it more likely to activate the existing protease. An extra level of control is thereby added, increasing the sensitivity of the cascade.[36]

The key to understanding the evolution of blood clotting is to appreciate that the current system did not evolve all at once. Like all biochemical systems, it evolved from genes and proteins that originally served different purposes. The powerful opportunistic pressures of natural selection progressively recruited one gene duplication after another, gradually fashioning a system in which high efficiencies of controlled blood clotting made the modern vertebrate circulatory system possible.

Can we know for sure that this is how blood clotting, or any other biochemical system, evolved? The strict answer, of course, is that we cannot. The best we can hope for from our vertebrate ancestors are fossils that preserve bits and pieces of their form and structure—and it might seem that their biochemistry would be lost forever. But that's not quite true. Today's organisms are the descendants of that biological (and biochemical) past, and they provide a perfect opportunity to test these ideas.

If the clotting cascade really evolved the way I have suggested, the clotting enzymes would have to be near-duplicates of a pancreatic enzyme and of each other. As it turns out, they are. Not only is thrombin homologous to trypsin,[37] a pancreatic serine protease, but the six clotting proteases (prothrombin and Factors VII, IX, X, XI, and XII, shown in Figure 5.3) share extensive homology as well. This is consistent with the notion that they were formed by gene duplication, just as suggested. But there is more to it than that. We could take one organism—humans, for example—and construct a branching tree based on the relative degrees of similarity and difference between each of the clotting proteases. Now, if the gene duplications that produced the clotting cascade occurred long ago in an ancestral vertebrate, we should be able to take any other vertebrate and construct a similar tree in which the relationships between the clotting proteases match the relationships between the human proteases. This is a powerful test for our scheme because it requires that sequences still undiscovered should match a particular pattern. And, as anyone knows who has followed the work in Doolittle's lab over the years, it is also a test that evolution passes in one organism after another.[38]

Many other tests and predictions can be imposed on the scheme as well, but one of the boldest was made by Doolittle himself more than a decade ago. If the modern fibrinogen gene really was recruited from a duplicated ancestral gene, one that had nothing to do with blood clotting, then we ought to be able to find a fibrinogen-like gene in an animal that does not possess the vertebrate clotting pathway. In other words, it ought to be possible to find a nonclotting fibrinogen protein in an invertebrate. Doolittle bet that it was possible. That was a mighty bold prediction, because if it could not be found, it would cast Doolittle's whole evolutionary scheme into doubt.

In 1990, Xun Yu and Doolittle won the bet, finding a fibrinogen-like sequence in the sea cucumber, an echinoderm.[39] The vertebrate fibrinogen gene, just like genes for the other proteins of the clotting sequence, was formed by the duplication and modification of preexisting genes.

Now, it would not be fair, just because we have presented a realistic evolutionary scheme supported by gene sequences from modern organisms, to suggest that we now know *exactly* how the clotting system has evolved. That would be making far too much of our limited ability to reconstruct the details of the past. Nonetheless, there is little doubt that we do know enough to develop a plausible—and scientifically valid—scenario for how it might have evolved. And that scenario makes specific predictions that can be tested and verified against the evidence.

A Dip in the Lobster Pot

A further example of evolution's remarkable ability to fashion new uses for existing proteins can be found in some of the crustaceans. Like us, crabs and lobsters have circulating proteins that will gel to form fibrous clots when blood vessels are broken. In fact, the superficial resemblances between crustacean and vertebrate clotting are so great that the lobster clotting protein has actually been given the same name as our own principal clotting protein, "fibrinogen." But lobster fibrinogen is an entirely different protein, and crustacean blood clots in an entirely different way. It turns out that a lobster clot is formed when an enzyme cross-links thousands of soluble fibrinogen molecules into a large, insoluble, flow-stopping clot. Where does the enzyme come from? The enzyme, technically known as a transglutaminase, is found inside most cells. When blood vessels are broken, cells are broken, too, and this enzyme (along with many others) leaks out. Since there's always plenty of lobster fibrinogen in the blood, the enzyme goes to work right away, seals the clot, and enables the animal to survive. Next time you're picking out a live lobster for dinner, you may still want to avoid the one that's missing a claw, but at least now you know why he didn't bleed to death when he lost it.

Could this system have evolved? It does indeed fit the test of irreducible complexity—the cellular enzyme wouldn't be any good without the fibrinogen protein, and it would be pointless to have fibrinogen in the lobster's bloodstream without a cross-linking enzyme. Another problem for evolution?

Not once you begin to look into things. The chemical differences between crustacean and vertebrate fibrinogen were interesting enough to Russell Doolittle that his laboratory isolated the lobster clotting protein for analysis. As they suspected, the protein was not at all like vertebrate fibrinogen. Unexpectedly, it was very, very much like another protein, one called vitellogenin.

Lobsters, like other animals, lay eggs. In order to nourish their young, they pack those eggs with as much nutritional value as their bodies can spare, and vitellogenin is the key to how they deliver that nutrition. Cells throughout the body manufacture this large protein and release it into the bloodstream. The ovary takes up the protein, cuts it into smaller pieces, and converts it into the proteins that make up the yolk of the egg. This means that the circulatory systems of nearly all egg-laying animals, including birds, amphibians, worms, and lobsters, are loaded with vitellogenin. All of a sudden, the evolutionary origin of lobster fibrinogen becomes clear.

Lobster fibrinogen didn't have to evolve from scratch because it was *already* there in the form of large amounts of circulating vitellogenin protein. The cross-linking protein didn't have to evolve from scratch either, because cells already contain several transglutaminase enzymes. What happens next? The ancestral forms of these two preexisting proteins would already allow for limited clot formation. That immediately places some selective value on their ability to interact. What happened next is clear, and as Doolittle's research has shown, is a common event in the evolution of new proteins. A mistake in either DNA replication or gene recombination accidentally duplicated the vitellogenin gene, producing an extra copy. Gene duplications occur all the time, and they are often of little importance. In this case, the existence of an extra copy meant that a mutation favoring the clotting reaction could occur in one of the genes without interfering with the yolk-building function of the other. From that moment on, the two vitellogenin genes were headed in different directions—one selected on the basis of its nutritional utility, and the other selected on the basis of clot formation.

By the same token, duplicate transglutaminase enzymes would have been selected for their cross-linking abilities, leaving the original versions to carry out reactions in unbroken cells. Every step of this evolutionary pathway, from the presence of the original genes to the separation of function in duplicate genes, would be favored by the classic mechanisms of Darwinian natural selection.[40] Over time, the split between classic vitellogenin and the new fibrinogen gene becomes greater and greater, but the modified gene still retains enough of its ancestral sequence to be recognizable. Enter modern science, and what do we see? A specialized set of clot-forming proteins whose evolutionary history is revealed in the details of protein sequence.

Far from being a *problem* for evolution, blood clotting in these organisms is a remarkable demonstration of the way in which evolution duplicates and then remodels existing genes to produce novel functions. The construction of a two-component clotting system nicely demonstrates something else as well. It dashes any claim that there is only one, irreducibly complex, mechanism to clot blood.

Michael Behe's purported biochemical challenge to evolution rests on the assertion that Darwinian mechanisms are simply not adequate to explain the existence of complex biochemical machines. Not only is he wrong, he's wrong in a most spectacular way. The biochemical machines whose origins he finds so mysterious actually provide us with powerful and compelling examples of evolution in action. When we go to the trouble to open that black box, we find out once again that Darwin got it right.

Remember the Darwin quotation that Behe used to justify his attack on evolution?

> If it could be demonstrated that any complex organ existed, which could not possibly have been formed by numerous, successive, slight modifications, my theory would absolutely break down.

In light of the biochemical systems we have just explored, it's too bad that Behe didn't go to the trouble to print the sentence that followed:

> But I can find no such case.

Neither, I would add, has Michael Behe.

God the Mechanic

The final blow to Behe's thesis comes, ironically, from Behe himself. Displaying a generous circumspection, he waves off any conclusion that the design he says he has found at the level of the cell can be shown to be the work of God Himself. That would be asking too much of science. Given the inadequacies of evolution, he says, the best we can do is to conclude that life is the work of an "intelligent agent."[41] Maybe God, but maybe something or somebody else.

Despite this reticence, Behe clearly wants to make a case for God as the designer of the cell's complex biochemistry. In Behe's view, God is a mechanic whose craftsmanship can be observed directly in the machinery of the cell. Behe makes the fatal theoretical mistake of taking himself a little too seriously. If God the mechanic did all this marvelous work, then he must have done it at a particular time in natural history—and there's the problem.

The great explanatory power of evolution comes, at least in part, from its ability to account for the sweeping changes in life that have occurred throughout natural history. Perhaps just a bit envious of this, Behe decided that the work of the God the mechanic had to be inserted into natural history, too.

Having accepted the validity of the geological ages and the fossil record—an intelligent scientist can hardly do otherwise—yet claiming that the complex biochemical systems he extols were fashioned by an intelligent agent, Behe faced a dilemma. Would he put forward the same scenario proposed by Phillip Johnson—in which God becomes a magician who periodically inserts new creatures into the world? Or would he think of something else? When would his intelligent agent go to work, and what exactly would he do? What Behe had to say on the subject amazed me:

> Suppose that nearly four billion years ago the designer made the first cell, already containing all of the irreducibly complex biochemical systems discussed here and many others. (One can postulate that the designs for systems that were to be used later, such as blood clotting, were present but not "turned on." In present-day organisms plenty of genes are turned off for a while, sometimes for generations, to be turned on at a later time.)[42]

According to Behe, our master mechanic rolled up his sleeves, packed all of his sweat, craftsmanship, and biochemical skill into a single, ancient cell, and then let things roll.

This explains, of course, why we don't observe biochemical miracles anymore. It also means that 4 billion years ago a humble bacterium was packed with genes that would be turned off for billions of years before they would be used to make the eukaryotic cilium, not to mention genes for blood clotting that would pass another billion inactive years in genetic cold storage before blood itself even existed. Carrying around all of the preformed genetic information required to make a human being, this bacterium must have been quite a sight to behold. It would have had to carry more than a thousand times the genetic information of one of today's bacteria which, strangely, bear no trace of the genes that once were "present," waiting to be "turned on" in the distant future. Behe does not regard this as a minor point—in fact, he even speculates that because of time travel, "biochemists in the future will send back cells to the early earth that contain the information for the irreducibly complex structures we observe today."[43]

If we choose to give Behe's theory serious consideration, if we treat it as a scientific hypothesis, then we are obliged to ask what would happen to those preformed genes during the billions of years to follow? As any student of biology will tell you, because those genes are not expressed, natural selection would not be able to weed out genetic mistakes. Mutations would accumulate in these genes at breathtaking rates, rendering them hopelessly changed and inoperative hundreds of millions of years before Behe says that they will be needed. As he should know, this is exactly what happens in controlled experiments when genes are silenced for long periods of time. This means, in genetic terms, that his idea of a designer doing all his work on that "first cell" is pure and simple fantasy.

What this colossal mistake shows is that one cannot develop a narrowly focused anti-evolution argument, like Behe's irreducible complexity, without paying attention to its impact of the other lines of evidence, including the fossil record, that profoundly support evolution. And that evidence, which Behe makes clear he accepts scientifically, squeezes him into a nonsensical position. It forces him, for the sake of consistency, to cobble his acceptance of the earth's well-documented natural history into the idea of God as a biochemical mechanic. The result is an absolutely hopeless genetic fantasy of "preformed" genes waiting for the organisms that might need them to appear gradually—and the utter collapse of Behe's hoped-for biochemical challenge to evolution. As H. Allen Orr wrote in *The Boston Review,* "The latest attack on evolution is cleverly argued, biologically informed—and wrong."[44]

Quo Vadis

In the last three chapters, we have considered three broad challenges to evolution, and found each of them to be lacking. You might think, given the revolutionary character of science, that people like Johnson and Behe would be peppering the scientific press with well-considered papers against evolution, that they'd be speaking at scientific meetings, and challenging the dominance of evolution on its own turf. They aren't, and neither are their colleagues.

To many scientists, this alone speaks volumes. It certainly means that the anti-evolutionists are not eager to expose their ideas to a hail of well-informed criticism, and quite a few of my scientific colleagues would go further than that. Knowing that I have met—and even debated—many of these critics, they ask me if the anti-evolutionists are "for real." They wonder how anyone could accept such ideas in the face of so much evidence to the contrary, and they question the sincerity and the motives of those involved in the anti-evolution movement. The more suspicious among them even wonder about a conspiracy.

The conspiracy idea makes little sense, owing to the dramatic differences between the distinct schools of anti-evolutionism. Behe, for example, accepts the reality of evolutionary change to an extent that even his supporters find surprising. In a 1995 debate,[45] I presented him with molecular evidence indicating that humans and the great apes shared a recent, common ancestor, wondering how he would refute the obvious. Without skipping a beat, he pronounced the evidence to be convincing, and stated categorically that he had absolutely no problem with the common ancestry of humans and the great apes. Creationists around the room—who had viewed him as their new champion—were dismayed. Behe's views stand in opposition to those of Phillip Johnson, who rejects any notion of a common ancestry for humans and other animals; and in bold contradiction to young-earth creationists like Henry Morris and Duane Gish, who reject common ancestry altogether and maintain that all species were separately created.

Johnson and Behe also accept what geologists tell us about the age of the earth, what astronomers tell us about the age of the universe, and what paleontologists tell us about the sequential appearances of species in the fossil record. The young-earth creationists reject all of this, and they view such concessions as logically fatal to their cause. Hard to imagine these folks getting together to conspire about anything.

Nonetheless, one thing does unify them—evolution—even if they present mutually contradictory arguments against it. As a group, their opposition to evolution is strong, passionate, almost visceral in a way that most scientists cannot comprehend. Why is this? Is it because the critics of evolution aren't very bright? Is it because they are blinded by a fundamentalism so total that it makes them impervious to scientific fact? Or is it because Americans neither respect nor understand science, as a British colleague once told me. I believe that none of these things are true.

By and large, the critics of evolution are not cynical opportunists. They aren't stupid, and they certainly understand how strong the scientific evidence is against them. So, why do they oppose evolution with such passion and persistence? I think I know, and as we shall see in the next chapter, many of my scientific colleagues, so baffled at the strength and depth of anti-evolution feelings in the U. S., would be surprised to discover that they are themselves a large part of the reason why.

6

THE GODS OF DISBELIEF

From time to time, would-be intellectuals of various persuasions have thought themselves justified—maybe even clever—to characterize evolution as nothing more than a creation myth, as just one story among many that account for the presence of our species on this planet. Creation myths serve many purposes. They produce a sense of unity and purpose among believers, reassuring them of their place in the grand scheme of things.[1] They'd like to believe, and to persuade others, that the fabric of Darwinism is nothing more than a just-so story woven from the desperation of secular humanism. By now it should be clear that this analysis is just a bit too clever for its own good.

In the real world of science, in the hard-bitten realities of lab bench and field station, the intellectual triumph of Darwin's great idea is total. The paradigm of evolution succeeds every day as a hardworking theory that explains new data and new ideas from scores of fields. High-minded scholarship may treat evolution (and it should) as just another scientific idea that could someday be rejected on the basis of new data, but actual workers in the scientific enterprise have no such hesitation—they know that evolution works. It works as a continuing process, and it works as a historical framework that explains both present and past.

As the previous three chapters have shown, the scattered ideas that have been put forward to disprove evolution do nothing of the kind. They fail on direct scientific terms, and they fail most notably at a task for which evolution is especially persuasive—as a seamless explanatory integration of past and present. Evolution does not just provide a historical story of how we came to be—it also explains why we are the way we are.

In 1998, the National Academy of Sciences matter-of-factly described evolution as "the most important concept in modern biology."[2] Few biologists would disagree. In fact, many of my friends and colleagues in research have told me flat-out they just cannot believe that evolution is still regarded as controversial in some quarters. Yet it is. The very reason the National Academy felt it necessary to say anything at all about evolution was their concern that continued resistance to the teaching of evolution has "contributed to widespread misconceptions about the state of biological understanding and about what is and is not science."[3]

In response, National Academy commissioned a blue-ribbon committee to update public understanding. That committee produced what is, by any standard, a fine document summarizing the scientific status of evolution, and containing an excellent set of recommendations for the teaching of evolution.[4] They left no doubt on the key scientific issues:

> Compelling lines of evidence demonstrate beyond any reasonable doubt that evolution occurred as a historical process and continues today.[5]

Is there any evidence that might support the alternatives to evolution we have visited in the last three chapters? Clearly not.

> It is no longer possible to sustain scientifically the view that living things did not evolve from earlier forms or that the human species was not produced by the same evolutionary mechanisms that apply to the rest of the living world.[6]

Couldn't have said it better myself.

Now that the National Academy has spoken, can we look forward to a dramatic increase in public acceptance of evolution? I think not.

Whatever the scientific and logical virtues of their report, and they are considerable, the National Academy failed to address—and maybe even to understand—the two most important sources of creationism's enduring appeal. The first, not surprisingly, is religion itself. Religion's power to motivate and inspire remains undiminished, even in these modern times. The second is the reflexive hostility of so many within the scientific community to the goals, the achievements, and most especially to the culture of religion itself. This hostility, sometimes open, sometimes covert, sharpens the distinctions between religious and scientific cultures, produces an air of conflict between them, and dramatically increases the emotional attractiveness of a large number of anti-scientific ideas, including creationism.

In a subtle way, the National Academy misread the issue, because public acceptance of evolution—or any other scientific idea—doesn't turn on the logical weight of carefully considered scientific issues. It hinges instead on the *complete* effect that acceptance of an idea, a world view, a scientific principle, has on their own lives and *their* view of life itself. America has become the greatest scientific nation in the world, largely because the optimistic and iconoclastic character of science is a perfect fit for the open and vigorous social temperament of American society. In one sense, the popular culture embraces evolution just as it has the rest of science—witness the public fascination with fossils and dinosaurs and the extraordinary popularity of movies like *Jurassic Park,* based on the central premise of evolutionary change. But in another sense it rejects that premise soundly. Less than half of the U.S. public believes that humans evolved from an earlier species.[7] The reason, I would argue, is not because they aren't aware of strength of the scientific evidence behind it. Instead, it is because of a well-founded belief that the concept of evolution is used routinely, in the intellectual sense, to justify and advance a philosophical worldview that they regard as hostile and even alien to their lives and values.

The Bosom of Abraham

Presented modestly and accurately, evolution is still a simple scientific idea. It claims only that material causes, the laws of physics and chemistry as played out in living things, are sufficient to account for the history and complexity of life. If evolution is neither more nor less than this simple scientific idea, then why does it engender such hostility? There are many reasons, but the most important ones are historical.

In a time when the world was beyond human comprehension, it was only natural to attribute the wonders of nature to the power and magnificence of the Creator. While any religious person today, myself included, continues to believe in a Creator, at that time such faith included an affirmation that nature was, in whole and in part, beyond our understanding. By any measure, the great and lasting gift of science has been to lift the burden of that affirmation from human imagination.

We have been freed to understand the change of seasons not as divine whim but as a consequence of the tilt of the earth's axis in relation to its orbit around the sun. We watch the movement of tides under the calculable power of gravity, produce new substances by rearranging the atoms of raw materials, and exploit the energy of elementary particles to power our homes and send messages through space. We have learned enough of the natural world to understand that it operates according to physical principles that are accessible through science.

In a sense, all that Charles Darwin did was to extend this understanding, so clear and so powerful in the physical world, into the sphere of biology. Evolutionary biologist Douglas Futuyma echoed the sentiments of many scholars when he wrote:

> By coupling undirected, purposeless variation to the blind, uncaring process of natural selection, Darwin made theological or spiritual explanations of the life processes superfluous. Together with Marx's materialist theory of history and society and Freud's attribution of human behavior to influences over which we have little control, Darwin's theory of evolution was a crucial plank in the platform of mechanism and materialism—of much of science, in short—that has since been the stage of most Western thought.[8]

Science, by this analysis, is mechanism and materialism. And all that Darwin did was to show that mechanism and materialism applied to biology, too. As I hope the preceding chapters have shown, there is something very right about this. The application of materialist science has transformed biology from the observational discipline of the nineteenth century into a predictive and manipulative science that enters the twenty-first century as the most dynamic and far-reaching of all the scientific disciplines. Materialism in biology, the search for mechanistic and chemical explanations for living phenomena, works, and it works brilliantly.

Evolution dashed the hopes of those who might have seen life as the one thing in the natural world that science would *never* be able to explain. One by one, the physical objects of our daily existence, even the stars and sun, yielded to material explanations of their powers and origins. Maybe life would resist this trend. Maybe life would defy material explanation, would require a vital spirit, a divine spark that could never be held by the crude hands of the scientist. It didn't turn out that way. Life turned out to be chemical after all, bridging the final gap between biology and the physical sciences when James Watson, Francis Crick, and Rosalind Franklin elucidated the double-helical structure of DNA, the molecule of heredity.

In an important way, the modern science of molecular biology is the child of what Darwin began, an ultimate unification of biology with other sciences, all of which now seek to describe nature in material terms. So, how does this threaten religion?

In a strict and scientific sense, it doesn't. And I find it puzzling and disappointing that so many would have pinned their religious hopes on the *inability* of science to explain the natural world. In fact, I will argue later that an accurate and complete understanding of that world, even in purely material terms, should deepen and strengthen the faith of any religious person.

The National Academy of Sciences clearly endorses the view that science and religion need not be in conflict. Responding to the question of whether or not a person can believe in God and still accept evolution, their report asserted that many people do exactly that:

> At the root of the apparent conflict between some religions and evolution is a misunderstanding of the critical difference between religious and scientific ways of knowing. Religions and science answer different questions about the world. Whether there is a purpose to the universe or a purpose for human existence are not questions for science. Religious and scientific ways of knowing have played, and will continue to play, significant roles in human history.[9]

I did not participate in the drafting of this careful language, although I certainly approve of these sentiments. There is indeed a degree of separation between religion and science, and it is important to point out that science and religion, in many respects, address quite different issues about the nature of reality.

A few years ago, no less an evolutionist than Stephen Jay Gould seemed to argue that science and religion occupy separate, but equally important, places in the intellectual world. In March of 1997, Gould devoted his monthly column in *Natural History* to a glowing analysis of the extraordinary letter that Pope John Paul II had written in 1996 to the Pontifical Academy of Sciences. The papal letter made it clear that there was no essential conflict between biological evolution and the world's largest Christian faith. Gould wrote:

> Science and religion are not in conflict, for their teachings occupy distinctly different domains. . . .
>
> The lack of conflict between science and religion arises from a lack of overlap between their respective domains of professional expertise—science in the empirical constitution of the universe, and religion in the search for proper ethical values and the spiritual meaning of our lives. The attainment of wisdom in a full life requires extensive attention to both domains.[10]

Gould's transcendent prose reserves a place of respect for religious belief, and more than once I have pointed to this exceptional essay as proof that evolution is not inherently hostile to religious belief.

The opponents of evolution are not at all convinced by these fine words. In many respects they regard them as nothing more than a smoke screen, a disarming party line intended to mute the opposition to evolution. Some wonder if Gould, in his heart, really believes these words. Late in 1997, Phillip Johnson described Gould's essay as "a tissue of half-truths aimed at putting the religious people to sleep, or luring them into a 'dialogue' on terms set by the materialists."[11] Had Johnson seen Gould on television a year later, his sense of Gould's duplicity might have been dramatically confirmed:

> INTERVIEWER: Gould disputes the religious claim that man is at the center of the universe. The idea of a science-religious dialogue, he says, is "sweet" but unhelpful.
>
> [Speaking to Gould]: Why is it sweet?
>
> GOULD: Because it gives comfort to many people. I think that notion that we are all in the bosom of Abraham or are in God's embracing love is—look, it's a tough life and if you can delude yourself into thinking that there's all some warm and fuzzy meaning to it all, it's enormously comforting. But I do think it's just a story we tell ourselves.[12]

Hard to see how something Gould regards as "just a story we tell ourselves" could also be an obligatory step in "the attainment of wisdom."

Other proponents of evolution would be neither so vague nor so genteel. All scientists enjoy interpreting their discoveries to the public, and sometimes this means commenting upon the meaning of scientific ideas. Evolutionary biology has drawn more than its share of scientific writers willing to tell an eager public what its discoveries mean. Many, if not most, are frankly inclined to use evolution as a weapon against religion, agreeing with the sentiments of Richard Dawkins—who, as we have seen, counted among Darwin's achievements the fact that he "made it possible to be an intellectually fulfilled atheist."[13]

Dawkins has actually gone much farther than that. The world revealed by evolution, he tells us, is not at all the kind of world we would expect of one that was subject to the designs of a Creator. Instead, a universe ruled by materialist evolution

> would be neither evil nor good in intention. It would manifest no intentions of any kind. In a universe of physical forces and genetic replication, some people are going to get hurt, other people are going to get lucky, and you won't find any rhyme or reason in it, nor any justice. The universe that we observe has precisely the properties we should expect if there is, at bottom, no design, no purpose, no evil and no good, nothing but blind, pitiless indifference.[14]

Nothing but "blind, pitiless indifference?" Little wonder that people who see the world as a place of deliberate moral choice, who see clear differences between good and evil, and who cherish virtues such as courage, honesty, and truthfulness would take issue with that statement. Since such characterizations are presented as the direct implication of evolutionary theory, one might fairly conclude from them that evolution itself is their enemy.

If Dawkins left even a sliver of doubt as to the anti-religious implications of evolution, William Provine, biologist and historian of science at Cornell, helped to remove it:

> Modern science directly implies that there are no inherent moral or ethical laws, no absolute guiding principles for human society. . . . We must conclude that when we die, we die, and that is the end of us. . . . Finally, free will as it is traditionally conceived—the freedom to make uncoerced and unpredictable choices among alternative courses of action—simply does not exist. . . . There is no way that the evolutionary process as cur-

rently conceived can produce a being that is truly free to make moral choices.[15]

Statements like these naturally provoke powerful reactions. One such reaction, unfortunately, would be to take at face value the Provine and Dawkins claim that evolution excludes God, and then move to protect religion by rejecting naturalism completely.

William A. Dembski, a philosopher at the Discovery Institute who has championed intelligent design, has done exactly that: "As Christians we *know* that naturalism is false. Nature is not self-sufficient."[16] Somehow, in casting "naturalism" as an intellectual devil, Dembski concludes that Christians *may* accept material explanations for the change of seasons and the light of the sun, but they apparently *may not* do so for the origin of species. How he makes this conclusion is beyond me, but his motivation for doing so is clear. He has seen the likes of Dawkins and Provine hone a fine edge on evolution to craft it into an anti-religious weapon, and he is determined to resist the use of that weapon at all costs.

Are such opponents of evolution sincere? Several years ago, I was invited to Tampa, Florida, to debate the issue of evolution with Henry Morris, founder of the Institute for Creation Research and one of the most influential of the young-earth creationists. The debate had been occasioned by the passage of a curriculum mandating the inclusion of so-called creation science in high school biology. In front of a large audience, I hammered Morris repeatedly with the many errors of "flood geology" and did my best to show the enormous weight of scientific evidence behind evolution. One never knows how such a debate goes, but the local science teachers in attendance were jubilant that I scored a scientific victory.[17]

As luck would have it, the organizers of this event had booked rooms for both Dr. Morris and myself in a local motel. When I walked into the coffee shop the next morning, I noticed Morris at a table by himself finishing breakfast. Flushed with confidence from the debate, I asked if I might join him. The elderly Morris was a bit shaken, but he agreed. I ordered a nice breakfast, and then got right to the point. "Do you actually believe all this stuff?"

I suppose I might have expected a wink and a nod. We had both been paid for our debate appearances, and perhaps I expected him to acknowledge that he made a pretty good living from the creation business. He did nothing of the sort. Henry Morris made it clear to me that he believed *everything* he had said the night before. "But Dr. Morris, so much of what you argued is wrong, starting with the age of the earth!" Morris had been unable to answer the geological data on the earth's age I had presented the night before, and it had badly damaged his credibility with the audience. Nonetheless, he looked me straight in the eyes. "Ken, you're intelligent, you're well-meaning, and you're energetic. But you are also young, and you don't realize what's at stake. In a question of such importance, scientific data aren't the ultimate authority. Even you know that science is wrong sometimes."

Indeed I did. Morris continued so that I could get a feeling for what that ultimate authority was. "Scripture tells us what the right conclusion is. And if science, momentarily, doesn't agree with it, then we have to keep working until we get the right answer. But I have no doubts as to what that answer will be." Morris then excused himself, and I was left to ponder what he had said. I had sat down thinking the man a charlatan, but I left appreciating the depth, the power, and the sincerity of his convictions. Nonetheless, however one might admire Morris's strength of character, convictions that allow science to be bent beyond recognition are not merely unjustified—they are dangerous in the intellectual and even in the moral sense, because they corrupt and compromise the integrity of human reason.

My impromptu breakfast with Henry Morris taught me an important lesson—the appeal of creationism is emotional, not scientific. I might be able to lay out graphs and charts and diagrams, to cite laboratory experiments and field observations, to describe the details of one evolutionary sequence after another, but to the true believers of creationism, these would all be sound and fury, signifying nothing. The truth would always be somewhere else.

In 1996, Jack Hitt wrote an article for *Harper's* magazine, "On Earth as It Is in Heaven," in which he captured this feeling exactly. Describing a series of "field trips" in which he dutifully recorded the curious geology taught by a succession of creationist professors, at last he came upon Kurt Wise, a professor of science at Bryan College in Tennessee. Wise's Ph.D. is from Harvard University, where his doctoral advisor was none other than Stephen Jay Gould. Despite his scientific pedigree, Wise is a committed and articulate young-earth creationist, and Hitt was clearly taken by the depth of Wise's convictions.

After listening to the stories and constructs of creationist geology, Hitt found himself comparing them favorably to the, in his words, "random evolution, meaningless mutations, trial and error (mostly error), aimless procreation, the pointless void of space, the cold materialism of Darwin's damn theory." On every intellectual level, as Hitt made clear, evolution holds the scientific cards. Yet, when he considered what it would feel like to embrace the creationism presented by Wise, it produced an emotional tide powerful enough to sweep Darwin right off the table:

> I felt again the warmth of believing that for every inch of infinity there has already been an accounting. Everything has a reason for being where it is. . . . I had felt it before, in childhood, when everything around me radiated with specific meaning and parental clarity. That, after all, is what creationists feel that evolution has stolen from them.[18]

To Wise and many others, the disciples of evolution have crushed the innocence of childhood, poisoned the garden of belief, and replaced both with a calculating reality that chills and hardens the soul. How sweet it would be to close one's eyes to "Darwin's damn theory," and once again sleep blissfully (Gould notwithstanding) in the bosom of Abraham.

A Dangerous Idea

At its heart, evolution is a modest idea, a minimal concept, just two points, really. First, the roots of the present are found in the past; and second, natural processes, observable today, fully explain the biological connections between present and past. On purely scientific terms, those two points leave very little to argue about.

But unlike mitosis, protein synthesis, or signal transduction, evolution almost begs for extrapolation. The success of the natural sciences have led one analyst after another to extend Darwinian thinking into a series of distinctly nonbiological enterprises, even into the study of religion.

Notorious among these was the social Darwinist movement of the late nineteenth century. Comparing social and political units to living organisms, Herbert Spencer argued that the state should not interfere with the social equivalent of natural selection. This meant, according to

Spencer, that aid to the poor, universal education, and laws regulating factory working conditions were all bad ideas because they might interfere with the natural order of social competition. Not surprisingly, these views led Spencer to become an outspoken opponent of socialism who counted American capitalist Henry Ford among his admirers.[19]

Ironically, many of the proponents of socialism also declared their allegiance to Darwin. Speaking at Karl Marx's funeral in 1883, Friedrich Engels proclaimed, "Just as Darwin discovered the law of development of organic nature, so Marx discovered the law of development of human history."[20] Ten years earlier, Marx had sent Darwin a new edition of *Das Kapital,* inscribed as a gift from a "sincere admirer."[21]

For a biologist, the embrace of Darwinism by both capitalism and socialism is easy to understand. Evolution is good science, and ideological partisans, even contradictory ones, seek to bolster their causes by associating with it. The truth is that evolution, however persuasive, is a *biological* theory fashioned to explain descent with modification. By what logic could anyone pretend that principles found to apply biological systems must also apply to social organizations, societies, or nations? Economic theorists may derive colorful analogies, insight, and even inspiration from evolution, but to prove the validity of economic theories by mere reference to evolution, a biological theory, seems to me to be stretching Darwin's good work too far.

I raise this point not to pick on economics, but to illustrate how easily evolution spills out of biology into other disciplines. The logic of Daniel C. Dennett's 1995 book, *Darwin's Dangerous Idea*, is a perfect example of what I mean. Dennett argues that Darwin deserves an award "for the single best idea that anyone has ever had."[22] It's a sentiment I share. Dennett is a philosopher who fully appreciates the explanatory power of evolutionary theory in accounting for the complexity and apparent design of complex biological systems. He also regards Darwin's idea as a kind of "universal acid," an intellectual substance so powerful that it can dissolve its way right out of the discipline of biology into fields as distant as moral reasoning and cosmology. In each case, according to Dennett, Darwinism is dangerous because of the threat it poses to nonmaterialist value systems. In the end, Darwinism can be expected to revolutionize thinking in psychology, ethics, art, politics, and every conceivable aspect of human culture.

Dennett is particularly eager to invoke Darwinian mechanisms to account for the development of the human mind:

> If mindless evolution could account for the breathtakingly clever artifacts of the biosphere, how could the products of our own "real" minds be exempt from an evolutionary explanation?[23]

In general, I agree. Although there is no reason to maintain that evolution has not shaped the human brain in the same way as it has shaped the rest of our bodies, this proposition has led him into a bitter academic food fight[24] on the issue of human language. The controversy illustrates what happens when evolutionary explanations are pressed beyond their scientific limits.

We share wide-ranging physical similarities with our nearest evolutionary relatives, the great apes. These similarities led Darwin to suggest that Africa, where gorillas and chimpanzees live today, would be the best place to search for our common ancestors. The suggestion has resulted in a rich harvest of hominid fossils. Molecular genetics has underscored this relationship, showing clearly that we share nearly all of our genetic information with our primate kin. These animals, as intelligent and remarkable as they are, clearly do not share our capacity for language. True, patient and sometimes controversial studies have shown that gorillas and chimps can be taught to manipulate a handful of symbols, articulate a few words, and maybe engage in a little bit of abstract reasoning. Nonetheless, even the most generous analysis of language development in these organisms shows an enormous gulf separating our verbal abilities from theirs.

Neural and cognitive scientists assert, on the basis of strong physiological evidence, that human language ability is wired into the brain. We can acquire and manipulate language so well because the generous circuitry of our brains gives us the capacity to do so. The next question, of course, is how the brain got that way. Whatever the answer to this question might be, it will have to be a corollary of a larger, more basic one: How did natural selection act over the past 4 million years to enlarge the average size of the australopithecine brain from just over 400 cubic centimeters to the present human average of 1500 cubic centimeters?

A number of scientists have argued that the capacity for human language is what Stephen Jay Gould calls a "spandrel," an *artifact* of an increase in brain size actually driven by selective forces acting on factors other than language itself. An evolutionary spandrel, in Gould's rich and expansive vocabulary, is an unexpected structure or capacity that appears as the result of natural selection. He derives the term by reference to the magnificent Church of San Marco in Venice, whose great ceilings arch downward onto supporting columns. Where these arches merge into the columns, they form tapering triangular spaces, *spandrels,* beautifully decorated by San Marco's artists and craftsmen. As Gould and geneticist Richard Lewontin pointed out in a landmark 1978 paper,[25] an uncritical visitor might quickly assume that the spandrels were an intentional architectural feature of the church, key surface spaces that the builders intended to create when they designed the structure. Nothing of the sort is true. The spandrels were a feature coincidental to the architectural choice of circular arches to support the ceiling.

In Gould's view, our brain's remarkable capacity for language is just such an unexpected spandrel of the brain's enlargement the past few million years. In other words, natural selection supported larger brains for reasons that had nothing to do with language. Once the brain was large enough to make possible the creation and manipulation of language, true human speech appeared.

> Yes, the brain got big by natural selection . . . [but] the brain did not get big so that we could read or write or do arithmetic or chart the seasons—yet human culture, as we know it, depends upon skills of this kind. . . . [T]he universals of language are so different from anything else in nature, and so quirky in their structure, that origin as a side consequence of the brain's enhanced capacity, rather than as a simple advance in continuity from ancestral grunts and gestures, seems indicated.[26]

This means, of course, that Gould does not view language itself as one of the adaptations favored by natural selection. Dennett does not agree, and attacks this view with every weapon at his disposal. As one of his main authorities, he uses the work of MIT scientist Steven Pinker, who with Paul Bloom has written: "[W]e conclude that there is every reason to believe that a specialization for grammar evolved by a conventional neo-Darwinian process."[27] Dennett endorses their analysis, in opposition to Gould:

> They [Pinker and Bloom] arrived at this . . . conclusion by a patient evaluation of multifarious phenomena that show beyond a reasonable doubt—surprise, surprise—that the "language organ" must indeed have evolved many of its most interesting properties as adaptations, just as any neo-Darwinian would expect.[28]

The stakes in this little debate are much greater than they may seem—at issue is nothing less than the independence of human thought and reason. To Gould, language and the mind may exist *because* of natural selection, but were never the *objects* of selection. Therefore, any attempt to explain all of human nature by means of evolutionary advantage will fail. To Dennett, natural selection really does explain the whole of language, mind, and culture. Therefore, every aspect of human thought, reason, and emotion is a direct product of the forces of evolution that have shaped the human mind.

Since no one has yet been able to determine which of many possible adaptive traits caused natural selection to favor an increase in size of the *entire* human brain, on what basis does either side of this debate assert that they know anything about the forces that acted on just one part of it—the capacity for language? As an experimental scientist, I would carefully suggest that they know very little. They have been arguing about two perfectly plausible schemes for how language *might* have evolved, in the absence of enough data to assert with any certainty how it actually *did* evolve. Disputes like this do not necessarily lend a stamp of scientific fact to either viewpoint.

The explanations of both Gould and Dennett are perfectly reasonable speculations on the evolution of language. I differ with them both, and most especially with Dennett in his assertion that we can extend Darwinian analysis to solve problems in nearly any field in truly scientific fashion. Science works. And Darwinism is science. Therefore, he reasons, a Darwinian analysis of *any* discipline is also a scientific one. If that were true, Darwin's great idea would be even more than the "greedy reductionism" he condemns in his book. It would lead to the conclusion that any trait, any character, and any behavior of any organism must be the direct product of natural selection. No spandrels allowed, and no need, apparently, to wait for experimental evidence.

Dennett's views, widely shared by social psychologists and behavioral ecologists, would allow every aspect of human thought to be explained away as an evolutionary adaptation. As we have just seen, this applies to the capacity for language, but just as surely it can be applied to the capacity for religion. As a result, his view of evolution leaves no room for God, except as a psychological curiosity to be studied and explained. Our abilities to imagine the divine, which are surely part of human nature, must exist because of natural selection. They surely do not exist because the Deity is real.

By such logic, one begins with Darwin's great idea, and fashions from it an ultimate weapon for the struggle against spirituality. Dennett makes clear his intentions to use that weapon, his universal acid, to dissolve the hold that religion has on human thought:

> What, then, of all the glories of our religious traditions? They should certainly be preserved, as should the languages, the art, the costumes, the rituals, the monuments. Zoos are now more or less seen as second class havens for endangered species, but at least they are havens, and what they preserve is irreplaceable. The same is true of complex memes [religion] and their phenotypic expressions [churches]. . . .[29]
>
> My own spirit recoils from a God Who is He or She in the same way my heart sinks when I see a lion pacing neurotically back and forth in a small zoo cage. I know, I know, the lion is beautiful but dangerous; if you let the lion roam free, it would kill me; safety demands that it be put in a cage. Safety demands that religions be put in cages, too—when absolutely necessary.[30]

Although Dennett sees much value in the artistic, literary, and even the social works of religion, his words make it clear that one of the hopes he holds for Darwin's dangerous idea is that it will eventually put religious beliefs safely behind bars. A believer is left to hope that the cages Dennett imagines are only intellectual constructs, but one never knows for sure.

On Human Nature

One of the boldest extensions of Darwinian analysis has come from a biologist whose specialty is the study of behavior. This is not surprising for the simple reason that behavior, in the broadest sense, is everything that people do. Art, music, literature, and even science are all forms of human behavior. Throughout the ages, our ability to produce these elements of high culture has been taken as evidence of the specialness of human nature. Such behaviors are the very things that distinguish us from the animals, that make us human, and can be used to mark us as the children of God.

Edward O. Wilson is a Harvard University biologist whose own research deals with ants, wasps, and bees, a fascinating group of animals known as the social insects. Wilson and his associates have studied insect societies for years, and we owe a great deal of our understanding of these remarkable organisms to him. Wilson saw the exquisite behavior patterns of the social insects as a special case of inherited behavior, and coined the term "sociobiology" to refer to the biological basis of social behavior. Wilson is a brilliant man who has labored for many years to achieve a synthesis between his biological analysis of behavior and other forms of human endeavor.

On Human Nature, Wilson's 1978 Pulitzer Prize–winning book, applied a Darwinian analysis of human behavior to a variety of social institutions. It earned him praise, but no small amount of criticism, especially from feminists, who saw his biologically based arguments as nothing more than sexist apologetics for an oppressive status quo. Liberal social theorists also were bothered by what they saw as his biological defenses of social inequality, and by his descriptions of Marxism as badly flawed and "mortally threatened" by sociobiology. Some of Wilson's sharpest barbs were reserved for religion. In his view, Darwinism had provided nearly enough information to exclude the existence of God.

> If humankind evolved by Darwinian natural selection, genetic chance and environmental necessity, not God, made the species. Deity can still be sought in the origin of the ultimate units of matter, in quarks and electron shells (Hans Küng was right to ask atheists why there is something instead of nothing) but not in the origin of species. However much we embellish that stark conclusion with metaphor and imagery, it remains the philosophical legacy of the last century of scientific research.[31]

If science can exclude, or almost exclude, the Deity, then where did religious belief come from? The core idea of sociobiology is that genetically determined behaviors are really just biological traits. This means that natural selection can act on those behaviors just as surely as it can act on the shape of a wing or the color of fur. Those behaviors that are most favorable, that best aid their possessors in the struggle for survival and reproductive success are, naturally enough, the ones that endure. This allows us to understand why a female bird would give food to her babies even though it means reducing the food that is available for *her own* survival. If that food-giving behavior is programmed by genes (and it probably is), then there a good chance that these genes are helping copies of themselves to survive. How can this be true? Any gene present in the mother has a fifty-fifty chance of being present in any one of the chicks. So a gene that induces the feeding of offspring, which we might consider a fine and noble thing to do, is actually a "selfish" gene, that programs a behavior to help copies of itself (in the babies) to survive.[32]

This kind of analysis can be applied to the religious impulse, too. Noting that religions are widespread throughout the world, however much their specific rituals and traditions vary, Wilson concludes that the religious impulse is a universal aspect of human nature. I certainly agree, and so would most sociologists and anthropologists. Wilson then steps back, as a biologist used to analyzing insect societies might, and asks what the adaptive significance of religious behavior might be.

> The highest forms of religious practice, when examined more closely, can be seen to confer biological advantage. Above all they congeal identity. In the midst of the chaotic and potentially disorienting experiences each person undergoes daily, religion classifies him, provides him with unquestioned membership in a group claiming great powers, and by this means gives him a driving purpose in life compatible with his self-interest.[33]

In Wilson's view, it is essential to remember that we humans, even at our most primitive, are social animals. This means that any gene programming a behavior to make one small group or tribe more cohesive than another might be favored by natural selection. A band of slightly religious hunter-gatherers might be just a little bit better in hunting and gathering than one that was less cohesive. To put it another way:

> When the gods are served, the Darwinian fitness of the members of the tribe is the ultimate unrecognized beneficiary.[34]

If this is true, as Wilson believes, then it is even possible to develop a Darwinian critique of social inequalities that have religious roots.

> Consequently religions are like other human institutions in that they evolve in directions that enhance the welfare of the practitioners. Because this demographic benefit must accrue to the group as a whole, it can be gained partly by altruism and partly by exploitation, with certain sectors profiting at the expense of others.[35]

There are problems, big problems, with this analysis. Not the least of these are the grand dimensions of Wilson's extrapolation from sociobiology to human society. Starting with biological fact—that all social behaviors in an organism (like an insect) that cannot learn must have a genetic basis—he leaps to conclusions that infer genetic causality to nearly all behaviors in humans, the very creatures that have taken learned behavior to new heights. At the very least, this kind of reasoning would allow Wilson to assign a genetic basis for *any* behavior observed in human society. While this is no doubt true for *some* behaviors, the human ability to generate culture, tradition, and language makes this a problematic claim, to state it kindly.

On just this score, Wilson has been roundly criticized. To be sure, he has agreed that culture and learning make a difference, but even on this point he is unwilling to concede that the difference could be very large.

> The genes hold culture on a leash. The leash is very long, but inevitably values will be constrained in accordance with their effects on the human gene pool.[36]

In other words, even cultural values are subject to natural selection, making them the products of evolution rather than of human consciousness. Not even Daniel Dennett was willing to go quite this far, especially with respect to religion.

> Long before there was science or even philosophy, there were religions. They have served many purposes (it would be a mistake of greedy reductionism to look for a single purpose, a single *summum bonum* which they have all directly or indirectly served).[37]

Wilson asserts that *he* has found the *summum bonum,* the highest good, the ultimate goal that religion serves. It cements the cohesiveness of the tribe, ensuring the survival of the group. The success of evolution in explaining how social behaviors *can* be the objects of natural selection has led Wilson to a hubris in which all social behaviors are therefore *presumed* to be the results of natural selection. It's important to note that this could only be the case if selection could have acted directly upon each social behavior, including religion. Despite Wilson's willingness to construct Paleolithic stories of religious cohesion favored by group selection, there is little genuine evidence to support that claim.

Edward Wilson is a great scientist whose contributions to the study of behavior have earned him a lasting place in the scientific pantheon. He is kind and generous, an inspiring teacher and a supportive mentor. I have nothing but admiration for the man and his scientific work, and most especially for the depth and imagination of his writings on nature. Nonetheless, if opponents of evolution wanted to point to the works of just one biologist to argue that evolution is inherently hostile to religion, they would be hard-pressed to make a better choice than Ed Wilson.

Not only does Wilson dispute the validity of religious teachings, morals, and institutions, he also contends that his explanation for the existence of the religious impulse is the death knell for belief.

> We have come to the crucial stage in the history of biology when religion itself is subject to the explanations of the natural sciences. As I have tried to show, sociobiology can account for the very origin of mythology by the principle of natural selection acting on the genetically evolving material structure of the human brain. If this interpretation is correct, the final decisive edge enjoyed by scientific naturalism will come from its capacity to explain traditional religion, its chief competitor, as a wholly material phenomenon. Theology is not likely to survive as an independent intellectual discipline.[38]

To Wilson, once an evolutionary explanation for the existence of religion has been fashioned, the very idea of God is doomed.[39]

Wilson never asks if there might be another way to view the religious impulse, that even if it is more the product of genes than culture, it still is fair to ask whether or not those genes might be the way a Deity ensured His message found receptive ground.

Wilson's reasoning is clear. Step-by-step, he has used the tools of Darwinism to explain every human behavior, including our sense of the divine, as the product of blind, uncaring natural selection. All of a sudden, Gould's questionable truce between the nonoverlapping spheres of science and religion has come completely unglued. In Edward O. Wilson's hands, Darwin's idea has become dangerous indeed.

The Fabric of Disbelief

I'd love to be able to claim that the anti-religious writings of Wilson, Dennett, and Dawkins, as well as the more refined slights of Gould, were aberrations. These folks are supposed to be scientists, and one might think that science—dealing with the material—should have nothing to say about religion—which deals with the spiritual. Their personal views on religion are just that—individual opinions on questions of faith that reside outside the sphere of science. But the reality of academic life is different.

Western intellectual life is tolerant almost by reflex. The academic establishment recoils instinctively at overt expressions of racism or anti-Semitism, and reacts with solemn outrage at anything that can be identified as prejudice or bias. On the surface, this tolerance extends to religion. I have never seen an American university that failed to make clear to prospective students, benefactors, or granting agencies just how open and tolerant a place it really was. But this happy prospect falls well short of complete embrace, especially when the issue is religion.

The problem is not intolerance in the old, ethnically driven sense. It is easy to get by on most American campuses identifying oneself as an ethnic Jew, a Muslim, a Catholic, or even a Southern Baptist. Religion is studied carefully in history, sociology, psychology, and philosophy departments. The varieties of religious experience that individuals bring to academic life are treasured as gifts to scholarship and experience. The problems come when one attempts to take religion seriously, actually to *believe* the stuff. Academia just isn't prepared for that.

The conventions of academic life, almost universally, revolve around the assumption that religious belief is something that people grow out of as they become educated. The prospect of an educated person who sincerely believes in God, who prays and fasts, or who is naive enough to think that there is actually such a thing as sin, is just not taken seriously. There is, in essence, a fabric of disbelief enclosing the academic establishment. My colleagues do their best to be open, fair-minded, and tolerant. They practice the wonderful virtues of free inquiry and free expression. But their core beliefs do not allow them to accept religion as the intellectual equal of a well-informed atheistic materialism.

In practice, their exultation at seeing evolutionary biology successfully provide material explanations for the origins of species and the history of life leads to triumphant excess. Even though philosophical conclusions about meaning and purpose are generally thought to lie outside science, any number of self-assured scientists display no hesitation in claiming that evolutionary biology is capable of making a powerful and profound statement on the ultimate meaning of things. Writing in the journal *Nature,* biologist David Hull asserted that evolutionary biology could tell us about the nature of God Himself.

> Whatever the God implied by evolutionary theory and the data of natural history may be like, He is not the Protestant God of waste not, want not. He is also not a loving God who cares about His productions. He is not even the awful God portrayed in the book of Job. The God of the Galápagos is careless, wasteful, indifferent, almost diabolical. He is certainly not the sort of God to whom anyone would be inclined to pray.[40]

The veiled condescension Gould had displayed towards religion could well be shrugged off—it's just his personal opinion, and not one for which he claims the authority of science. But such is not the case for the others. Dawkins draws directly on evolution to say that life is without meaning. Wilson finds that evolution can explain God away as an artifact of sociobiology. And Dennett is ready to dig quarantine fences around zoos in which religion, held safely in check, can be appreciated from a distance.

All of these writers have gone well beyond any reasonable *scientific* conclusions that might emerge from evolutionary biology. Without saying so directly, they have embraced a brand of materialism that excludes from serious consideration any source of knowledge other than science. They are not alone. Harvard geneticist Richard Lewontin endorses this view in so many words, and explicitly proclaims that it should be part of a social program to transform human society. As he puts it, when science speaks to the masses,

> the primary problem is not to provide the public with the knowledge of how far it is to the nearest star and what genes are made of, for that vast project is, in its entirety, hopeless. Rather, the problem is to get them to reject irrational and supernatural explanations of the world, the demons that exist only in their imaginations, and to accept a social and intellectual apparatus, *Science*, as the only begetter of truth.[41]

Surely Lewontin means that science is the only begetter of material truth, or that science is the only reliable way to discover the workings of the physical universe, right? He means a little more than that:

> We take the side of science in spite of the patent absurdity of some of its constructs, in spite of its failure to fulfill many of its extravagant promises of health and life, in spite of the tolerance of the scientific community for unsubstantiated just-so stories, because we have a prior commitment, a commitment to materialism. It is not that the methods and institutions of science somehow compel us to accept a material explanation of the phenomenal world, but, on the contrary, that we are forced by our *a priori* adherence to material causes to create an apparatus of investigation and a set of concepts that produce material explanations, no matter how counter-intuitive, no matter how mystifying to the uninitiated. Moreover, that materialism is absolute, for we cannot allow a Divine Foot in the door.[42]

Richard Lewontin's eloquent hostility to anything connected to God rises naturally from his conception of science as "the only begetter of truth"—a conception, incidentally, that relies on a prior commitment to philosophical materialism. To Lewontin, science is in the midst of a "struggle for possession of public consciousness between material and mystical explanations of the world," a struggle against ignorance *and* spirituality that it cannot afford to lose.

A Righteous Reactionism

Confronted by intellectual aggressions as great as those of Lewontin, it is easy to see why religious people might conclude that evolution is a threat to everything they hold most dear—maybe even a threat to freedom of religion itself. If evolution leads logically to the exclusion of

God from a meaningless universe, then evolution must be fought at every opportunity. As distasteful as it might be to creationists to see themselves as intellectual reactionaries, given the overt hostility of the evolutionary establishment, their defensiveness is easy to understand.

Kurt Wise, Gould's own creationist Ph. D. student, has no doubts as to the stakes of the struggle with materialism, and his carefully chosen words in a review of his mentor's book *Wonderful Life* serve to define the terms of the battle. One of the central ideas of *Wonderful Life* is the historical contingency of the evolutionary process, a theme that Gould develops to highlight what he views as the wild improbability of humankind's presence on the earth. As Wise puts it:

> If, as Gould argues, the evolutionary tape were played again, human life would not be expected. In fact, even if it were replayed a million times or more, man would not be expected again.
>
> To conclude, as Gould does, that man is "a wildly improbable evolutionary event," "a detail, not a purpose," and "a cosmic accident" is disconcerting to some, but not to Gould. To him, release from any purpose is "exhilarating" as it also releases any responsibility to any other, "offering us maximum freedom to thrive, or to fail, in our chosen way." If ever evolutionary theory has been elaborated to the point of complete incompatibility with a Christian world view, it is by the pen of Stephen Jay Gould in this, his most recent tome.[43]

As Wise makes clear, he believes that the real danger of evolutionary biology to Christianity is not at all what most scientists might suspect. It is not that evolution's version of natural history threatens to unseat the central Biblical myths of unitary creation and the Flood. Rather, it is the chilling prospect that evolution might succeed in convincing humanity of the fundamental purposeless of life. Without purpose to the universe, there is no meaning, there are no absolutes, and there is no reason for existence.

To Bruce Chapman, former chief of the United States Census Bureau who serves as president of the Discovery Institute, theft of meaning perpetrated by evolution has resulted in much more than personal disillusionment to those who have accepted the scientific worldview. It has had practical consequences.

> It can be argued that materialism is a major source of the demoralization of the twentieth century. Materialism's explicit denial not just of design but of the possibility of scientific evidence for design has done untold damage to the normative legacy of Judeo-Christian ethics. A world without design is a world without inherent meaning. In such a world, to quote Yeats, "things fall apart; the center cannot hold."[44]

Why is this so important? What happens if "the center cannot hold?" To the critics of evolution, the likes of Lewontin, Dennett, and Dawkins inhabit a world of *absolute* materialism in which God is truly dead. That may be fine for members of the intellectual elite, but if ordinary people were to discover that the ethical and moral principles derived from religion were nothing more than a convenient social fiction, all hell might break loose. They might behave as if anything were permitted, and society would come apart in a flash.

This is exactly what Johnson, Chapman, and their colleagues fear has already begun—a moral dissolution caused by the uncontrolled leakage of Darwinism into "soft sciences" such as sociology and psychology. Also in the postscript to Dembski's *Mere Creation*, Chapman described a pivotal case in American law as emblematic of the sad consequences of Darwinism.

In 1925, just *before* his trip to Tennessee to argue in the Scopes trial (on the issue of evolution), Clarence Darrow defended accused murders Nathan Leopold and Richard Loeb. These rich, well-educated youths had murdered a fourteen-year-old boy, and eventually admitted they had done so just for the thrill of it. Darrow advised them to enter pleas of guilty, and then, in a closing statement lasting twelve hours over two days, argued brilliantly against the death penalty. The judge, John R. Caverly, may have been moved by Darrow's arguments. He sentenced the murderers to life in prison, commenting on his belief that to Leopold and Loeb, "the prolonged years of confinement may well be the severest form of retribution and expiation."

Chapman, however, looks beyond the issue of the death penalty to another important argument introduced by Darrow—namely, that a swirl of deterministic forces made Leopold and Loeb less than fully responsible for their crimes. It was to accomplish this feat, according to Chapman, "that Darrow first introduced Darwinian arguments into criminology."

> Brilliantly and successfully, Darrow, their attorney, argued against the death penalty, asserting that the "distressing and unfortunate homicide" happened mainly because a normal emotional life failed to 'evolve' properly in the defendants. Speaking of Richard Loeb, he demanded, "Is Dickie Loeb to blame because . . . of the infinite forces that were at work producing him ages before he was born . . . ? Is he to blame because his machine is imperfect?"
>
> Leopold and Loeb—Darrow repeatedly called them children—were really helpless agents of their genes. "Nature is strong and she is pitiless. She works in her own mysterious way, and we are her victims. We have not much to do with ourselves. Nature takes this job in hand, and we play our parts."[45]

It seems that Richard Dawkins wasn't the first one to find a "blind, pitiless indifference" in nature. Clarence Darrow got there first, and used that characterization to drive a historic wedge between behavior and personal responsibility. In Chapman's view, no doubt, the world has traveled a Darwin-inspired straight line of moral decline ever since. If evolution is capable of breaking the legal and moral ties between criminal behavior and the individual, then the very foundations of our society are at risk. To Johnson and other opponents of evolution, the real risk is that evolution tells people that God is dead. And if people were to believe that, they might indeed behave as if all is permitted. Social chaos would result—or *has resulted,* depending upon the degree of pessimism with which one views the present state of American society.

The depth and emotional strength of objections to evolution sometimes baffle biologists who are used to thinking of their work as objective and value-free. The backlash to evolution is a natural reaction to the ways in which evolution's most eloquent advocates have handled Darwin's great idea, distilling from the raw materials of biology an acid of hostility to anything and everything spiritual.

This clash of two cultures extends over a battle line encompassing every moral, ethical, and legal issue of modern life. The giddy irony of this situation is that intellectual opposites like Johnson and Lewontin actually find themselves in a symbiotic relationship—each insisting vigorously that evolution implies an absolute materialism that is *not* compatible with religion. This means, in a curious way, that each validates the most extreme viewpoints of the other.

Johnson's willingness to dismiss scientific evidence confirms Lewontin's fear that someone "who could believe in God could believe in anything." *Anything* clearly includes the nonsense of creationism, which one might think Lewontin would want to counter by a vigorous exposure to the sunlight of truth. But he writes that "it is not the truth that makes you free. It is your possession of the power to discover the truth."[46] Lewontin, the evolutionist, is not interested in universal understanding and enlightenment. What he's shooting for is a social order in which right-thinking people (like him) will hold the absolute reins of cultural and intellectual power. To Lewontin, direct enlightenment of the public is impossible.[47] We evolutionists simply have to grab hold of power and run things because history tells us we are right. His sentiments confirm the worst fears of the creationists—that evolution isn't really about science, but is instead an ideology of belief, power, and social control.

Evolution is not any of those things. It really is about science, and this means that the question of evolution is fundamentally different from issues of social and moral values. It means that by looking at nature, we can find out whether or not evolution took place, whether it continues today, and whether it serves as an adequate explanation of the range and diversity of life. As we have already seen, it does. So, what are we to make of the clash of values between the spiritualism of the creationists and the materialism of the evolutionists? Just this: that the conflict depends, as most intellectual struggles do, on an unspoken assumption shared by both sides. That assumption is, if the origins of living organisms can be explained in purely material terms, then the existence of God—at least any God worthy of the name—is disproved. What I propose we do next is to ask if that assumption is true.

In a scientific sense, it is certainly true that the world runs according to material rules, that we are material beings, and that our biology works by means of the laws of physics and chemistry. To all of this, evolution added one important fact—namely, that our biological *origins* are material as well. The triumphalists of materialism now act as though this last achievement is enough to exclude the spiritual. The reactionists of creationism respond in kind, tilting comically at evolution's windmill with every trick at their disposal.

But what if both sides are wrong?

Is it possible that the proud materialism of evolution is entirely compatible with a religious world of value and meaning? The vocal advocates of absolute materialism certainly don't think so. Richard Lewontin has written, "To appeal to an omnipotent deity is to allow that at any moment the regularities of nature may be ruptured, that miracles may happen."[48] Exactly true. Orderly science requires the regularity of nature, a willingness to exclude the supernatural from its ordinary affairs. I do not explain experimental results in my lab as the product of divine intervention, so why should we allow for the Deity's work in nature outside the laboratory walls? Makes sense.

But Lewontin has missed something, something important. What if the regularities of nature were fashioned in a way that they *themselves* allowed for the divine? What if the logical connections he makes between materialism and atheism are flawed? What if the very foundations that seem to lock evolution and religion into conflict were built upon suspect ground? In other words, what if the gods of disbelief were false?

7

BEYOND MATERIALISM

My ancestors knew the gods, and so did yours. For thousands of generations, these men and women, mothers and fathers of all our families, huddled in fear each time the sun god dipped below earth's cold horizon. They prayed as children, asking mercy from the gods of darkness, the demons of the night, imploring favors from the moon and stars. Gods of the hunt and field controlled their tables, and fortunes rose or sank according to the whims of fairies and devils. For proof of their existence, one had to look no farther than a flower or a snowflake, the passage of the seasons, the slow and steady movement of the stars across a nighttime sky. What human hand could weave a rainbow or fashion a sunrise? What voice could stay the coming of a storm or tell us how the eagle flies?

Our gods did magic. They did the work of nature, and they ruled the lives of man. They warmed us some days, and on others they made us shiver. They healed us when they wished, and other times they struck us down with sickness and death. Most of all, they filled a need that all men have, a need to see the world as sensible and complete. Gods filled the voids in nature we could not explain, and they made the world seem whole.

Then something happened. Something wonderful. A few of our ancestors began to learn the rules by which nature worked, and after a while, we no longer needed Apollo to pull the sun's chariot across the sky. We no longer asked Ceres to waken seeds from winter sleep. The movements of the sun and moon became part of a mechanism, a celestial machine in which each motion could be calculated and explained. The boldest among us, having weighed the sun and moon and plotted the orbits of the planets, took inventory of the substance of the earth. They harnessed the fearsome power of the thunderbolt, drew fuel from the ground itself, and built machines that turned nighttime into day. The gods had lost their power.

Gradually, humans took up the greatest challenge of all—we sought to understand life. We learned the causes of sickness, and with such knowledge conquered the very diseases that had slain so many and produced such fear. We discovered the units of life itself, learned to read and understand the language of inheritance, and even began to edit and change the code of life. How did we do all this? What changed a minor species of bipedal primates into the masters of the planet? In a word, we learned to explore nature in the systematic way we now call *science*.

To be sure, the gods of our ancestors left stories, and from such treasures we fashion the literature and art that adorn a hundred human cultures. We should always be grateful for those riches, but the gods themselves are gone, and we are no longer subject to their tyranny. We live in a world where the real dangers come not from demons and spirits, but from the awful consequences of our own excesses as a species. And serious though they are, very few of us would be so foolish as to exchange those worries for a regression to the abject darkness of the past.

Whether you look at human history as an optimist or a pessimist, whether you greet a new century with dread or with enthusiasm, there is no denying the transforming reach of science and technology on human affairs. We are much better off in the material sense as a result of science, but in many ways that is a minor consequence of the rise of scientific thinking. The more profound effects of science have been to change the ways in which we view the world around us. As Richard Dawkins has pointed out, originally the gods themselves were a kind of scientific theory, invented to explain the workings of nature. As humans began to find material explanations for ordinary events, the gods broke into retreat. And as they lost one battle after another, a pattern was set up. The gods fell backwards into ever more distant phenomena until finally, when all of nature seemed to yield, conventional wisdom might have said that the gods were finished. All of them.

Lighting a Candle

Why should this be so? By definition, a god is a nonmaterial being who transcends nature, so why should science, which deals only with the material world, have anything to say about whether or not a god exists? In the rigorous, logical sense, it shouldn't. But we are a practical species, interested in getting results. Humans like to feel that their beliefs have a link to reality, and here's where science has it all over religion:

> You can go to the witch doctor to lift the spell that causes your pernicious anemia, or you can take vitamin B12. If you want to save your child from polio, you can pray or you can innoculate. If you're interested in the sex of your unborn child, you can consult plumb-bob danglers all you want (left-right, a boy; forward-back, a girl—or maybe it's the other way around), but they'll be right, on average, only one time in two. If you want real accuracy (here, ninety-nine percent accuracy), try amniocentesis and sonograms. Try science.[1]

A cynical critique of science might regard it as nothing more than a new kind of magic, useful only when it works as well as it does in these examples. But there is something genuinely different about this particular brand of magic. Trying science makes sense because science comes with a track record. Science works because it is based on causality. Once you understand a process, even a complex one, you can reduce it to the mechanistic sum of its parts. Then, everything that happens becomes an obligatory outcome of how those components interact. It's just something that happens. No longer magic, but just a simple (and predictable) outcome. That, to paraphrase the title of Carl Sagan's final book, is why science serves "as a candle in the dark" in this "demon-haunted world."

We can light that candle to explain the commonplace activities of everyday life, by showing that an understandable, *material* mechanism is at work in each of them—in short, by showing that the phenomenon at hand is a property of the ordinary stuff of nature. *That* is the working assumption of materialism—namely that nature itself is where we can find the explanations for how things work. It is also the credo of science—making science, by definition, a form of practical, applied materialism.

A key to the success of scientific materialism, which soon became a unifying assumption, was the finding that actions taking place at each level of nature could be explained as the effects of actions at lower levels. Heat was molecular movement. Sound was molecular vibration. Light was a form of electromagnetic radiation. The blink of my eye could surely be explained in physical and electrical terms. Everything fit together perfectly. Each part, each level of nature was part of a machine with predictable and understandable actions.

Scientific materialism rules out the influence of the divine from a particular phenomenon by the application of what we might call "deterministic reductionism." We can exclude the spiritual as the immediate cause for any event in nature by showing how that event is determined in material terms. All the levels in nature connect according to well-defined rules.

As time has gone by, every suggestion of miracle has seemed to yield to science's analytic touch. Until finally, near the end of the nineteenth century, scientists could be confident that they had found a truly interlocking mechanism in nature. There were a couple of loose ends in physics—including a few curious measurements for the speed of light—but those seemed merely troublesome details. The infant science of biochemistry already suggested that the molecules in living things obeyed the same rules of chemistry as everything else, and evolution had extended materialism to the origins of species as well. As biology advanced, life would surely gain a material explanation for each of its remarkable properties—and so it has. The living cell works by purely material rules.

The most important conclusion from the success of scientific materialism was philosophical as much as it was scientific. It had shown us that nature was organized in a systematic, logical way.

Science had shown that material mechanisms, not spirits, were behind the reality of nature. It had found that each level of analysis was connected to ones above and below in the same way that the function of a clock is connected to the gears and shafts and springs within. And it had given mankind a new view of ourselves as material beings.

Could there be anything left for God to do?

The Architect

With measured concern, more than a few thinkers figured that there would always be at least one job left over for the Deity—and it was no small detail, at that. The Western concept of God has never been as a grab bag of little animist tricks, sparkling in the fire and dancing on the tides, weaving fairy nets in the morning mist. The Western Deity, God of the Jews, the Christians, and the Muslims, has always been regarded more as the architect of the universe than the magician of nature. In a way, that places Him above much of the materialist fray.

This very Western idea of God as supreme lawgiver and cosmic planner helped to give the scientific enterprise its start. Many Eastern religions take the view that reality is entirely subjective, and that man can never truly separate himself from the nature he wishes to understand. Whatever the contemplative value of these ideas, the ancient Eastern intellectual was thereby relieved of any feeling that the workings of nature might reflect the glories of the Lord. The Westerner was not, and this is one of the reasons we can say—despite the extraordinary technical prowess of many Eastern cultures—that true empirical, experimental science developed first in the West. Hindu philosophers were left to contemplate the ever-changing dance of life and time, while Western scholars, inspired by the one true God of Moses and Muhammad, developed algebra, calculated the movements of the stars, and explained the cycle of the seasons.

In the past, the idea that nature was a complete, functional, self-sufficient system was seldom thought to be an argument against the existence of God. Quite the contrary, it was regarded as proof of the wisdom and skill and care of that great architect. The heavens in all their regularity reflected the grandeur of the Lord. And scientific investigation was regarded as a fine and appropriate way to get closer to the Creator's ways. According to Newton, "This most beautiful system of the sun, planets, and comets, could only proceed from the counsel and dominion of an Intelligent and Powerful Being."[2] Carolus Linnaeus (Carl von Linné), the founder of the modern system of biological classification, put it differently (and somewhat less modestly): *"Deus creavit. Linnaeus disposuit."*[3] This God, the architect, was a grand fellow indeed.

At one time this deist view of God was sufficiently persuasive to claim many of the leading men of the American Revolution, including Jefferson and Franklin, among its adherents. Part of their attraction to deism, no doubt, was that it suggested a moral logic to the universe, a kind of natural law that today we see reflected in the stirring language of the Declaration of Independence and the logic of the American Constitution.

Whatever the political attractions of this viewpoint, there was no denying that the seamless linkage of events in strict causality left precious little for deism's architect to do in the present—except to be that grim and distant first cause, the great designer who plumbed the heavens and set the stars in motion. And who now, presumably, snoozes in retirement.

Even with such illustrious supporters, deism was not to last as a serious religious movement. Its first great flaw was religious. Whatever their differences, the three leading Western religions generally agree on the existence of an active, personal, involved Creator. The retired clockmaker envisioned by deism could hardly be more removed from these traditions, and each emphatically rejected such a passive God.

Deism's second flaw may well have been scientific. In its purest form, deism is satisfying only when we visualize a static, controlled, finished creation, a mechanism of precise laws and well-kept orbits. In the seventeenth and eighteenth centuries, that was pretty much what natural science said about the universe. But by the time the nineteenth century drew to a close, that view was gone forever. Not only had a generation of geologists shown that the earth itself was changeable, but Charles Darwin had established the mutability of species, and severed forever Linnaeus's direct connection between each living thing and the hand of God.

To use the lingo of the information age, the real problem with deism was the *synergy* between itself and the physical sciences. In a universe ruled according to Newtonian laws, a physics of precision would govern the movements of atoms as well as the movements of stars. And this meant, once the material basis of life had been established, that all events, past and future, could be reduced to the predictable meshwork of cosmic gears. The human role could involve nothing more than watching the pages of history turn inflexibly towards a preordained future. A deist lawgiver might have been great and wise, but if he was thought of in the context of the Newtonian clockwork, our lives would become about as interesting as billiard balls, each move governed only by the forces and collisions of particles already set in motion. Lucky for us, that depressing prospect would never prevail—because at the very beginning of the twentieth century the clockwork upon which it depended would be broken forever.

HOT METAL

If you have ever held a metal wire over a gas flame, you have borne witness to one of the great secrets of the universe. As the wire gets hotter, it begins to glow, to give off light. And the color of that light changes with temperature. A cooler wire gives off a reddish glow, while the hottest wires shine with a blue-white brilliance. What you are watching, as any high school physics student can tell you, is the transformation of one kind of energy (heat) into another (light). As the wire gets hotter and hotter, it gets brighter. That's because if there is more heat energy available, more light energy can be given off, which makes sense.

Why does the *color* of that light change with temperature? Throughout the nineteenth century, that deceptively simple question baffled the best minds of classical physics. As the wire gets hotter and hotter, the atoms within it move more rapidly. Maybe that causes the color (the wavelength) of the light to change? Well, that's true, but there's more to it. Every time classical physicists used their understanding of matter and energy to try to predict *exactly* which wavelengths of light should be given off by a hot wire, they got it wrong. At high temperatures, those classical predictions were dramatically wrong. Something didn't make sense.

Late in the year 1900, Max Planck, a German physicist, found a way to solve the problem. Physicists had always assumed that light, being a wave, could be emitted from an object at any wavelength and in any amount. Planck realized that it was exactly those assumptions that were getting them in trouble, so he thought of a quick fix. He proposed that light could only be released in little packets containing a precise amount of energy. He called these packets "quanta," from the Latin word for "amount." All of a sudden, everything fell into place. Theory matched experiment, and the color spectrum emitted from glowing metal could now be precisely predicted. Problem solved.

Well, sort of. Planck's solution was more than a little trick with numbers. It had an important physical implication. It meant for openers, that light, packed into those tiny quanta, actually behaved like a stream of particles. In 1900, no one, least of all Planck, thought that made much sense. He assumed that the quantum was just a mathematical convenience, and that the real story would emerge before long.

Just two years later, another physicist turned Planck's experiment around in a most revealing way. In 1902, Philipp von Lenard explored a phenomenon known as the photoelectric effect. When a beam of light strikes a metal plate in a vacuum, electrons are kicked out of the metal. Since light is a form of energy, it is only natural that it can knock things around, so the ejection of the electrons made perfect sense. Lenard set up a clever apparatus that enabled him to measure the energies of the electrons as they were ejected from the plate. He discovered, to his amazement, that the energies of the electrons had nothing to do with the intensity of the light. Twice as much light ejected twice as many electrons, but the energies of those electrons did not change. Incredibly, it was the color of the light—its wavelength—that determined the energies of the electrons.

Lenard's results kicked around for a couple of years until a young patent office clerk named Albert Einstein explained them in a brilliant paper that won him his first (and only) Nobel Prize. Seizing upon Planck's ideas, Einstein proposed that the energy to eject a single electron from the plate came from a single quantum of light. That's why a more intense light (more quanta) just ejects more electrons. But the energy in each of those packets, the quantum wallop if you will, is determined by the wavelength, the color, of the light. At a stroke, Einstein had shown that Planck's quanta were not just theoretical constructs. Light really could behave as if it were made of a stream of particles, today known as photons.

All of this might have been sensible and comforting were it not for the fact that light was *already* known to behave as if it were a wave! So many experiments already had shown that light could be diffracted, that light had a frequency and a wavelength, that light spread out like a wave on the surface of a pond. Could all those experiments be wrong? No, they were not. All of those experiments were right. Light was *both* a particle and a wave. It was both a continuous stream and a shower of discrete quantum packets. And that nonsensical result was just the beginning.

Classical physics had prepared everyone to think of physical events as governed by fixed laws, but the quantum revolution quickly destroyed this Newtonian certainty. An object as simple as a mirror can show us why. A household mirror reflects about ninety-five percent of light hitting it. The other five percent passes right through. As long as we think of light as a wave, a continuous stream of energy, it's easy to visualize ninety-five percent reflection. But photons are indivisible—each individual photon must either be reflected or pass through the surface of the mirror. That means that for one hundred photons fired at the surface, ninety-five will bounce off but five will pass right through.

If we fire a series of one hundred photons at the mirror, can we tell in advance which will be the five that are going to pass through? Absolutely not. All photons of a particular wavelength are identical; there is nothing to distinguish one from the other. If we rig up an experiment in which we fire a single photon at our mirror, we cannot predict in advance what will happen, no matter how precise our knowledge of the system might be. Most of the time, that photon is going to come bouncing off; but one time out of twenty, on average, it's going to go right through the mirror. There is nothing we can do, not even in principle, to figure out when that one chance in twenty is going to come up. It means that the outcome of each individual experiment is unpredictable *in principle.*

Any hopes that the strange uncertainty of quantum behavior would be confined to light were quickly dashed when it became clear that quantum theory had to be applied to explain the behavior of electrons, too. Electrons are one of the building blocks of atoms, essential for determining the chemical properties of matter. Electrons are also subject to quantum effects. Their behavior in any individual encounter, just like the photon fired at the mirror, cannot be predicted, not even in principle.

As we try to analyze the behavior of photons and subatomic particles, things get very strange. Werner Heisenberg formulated the essence of the problem as the "uncertainty principle," which states, basically, that we just can't know everything about a particle, no matter how hard we try. Our knowledge of the mass of a particle and our knowledge of its momentum are interrelated in a frustrating way. If we determine a particle's position with absolute precision, we will know *nothing* about its momentum. Likewise, if we contrive an experiment in which its momentum can be measured precisely, we will lose all information regarding its position.

What had begun as a tiny loose end, a strange little problem in the relationship between heat and light, now is understood to mean that nothing is quite the way it had once seemed. As Heisenberg himself wrote:

> I remember discussions . . . which went through many hours till very late at night and ended almost in despair, and when at the end of the discussion I went alone for a walk in the neighboring park, I repeated to myself again and again the question: "Can nature possibly be absurd as it seemed to us in these atomic experiments?"[4]

One hundred years after the discovery of the quantum, we can say that the answer is yes, that is *exactly* what nature is like. At its very core, in the midst of the ultimate constituents of matter and energy, the predictable causality that once formed the heart of classical physics breaks down.

An Uncertainty of Meaning

Does the quantum nature of submicroscopic reality mean anything to us "big" folks who inhabit the "real" world? It certainly does. A slew of popularizers have spared nothing in inspired efforts to clarify the meanings of these discoveries. They range from Heisenberg himself, who wrote *Physics and Philosophy* in 1958[5] to speculate on the deeper implications of his discoveries, to Fritjof Capra, whose *Tao of Physics*[6] explored a theme of parallels between modern physics and Eastern mysticism. For the interested reader, few subjects in contemporary science—except, perhaps, for evolution—have been blessed by so many pages of explanatory prose.[7]

For present purposes, I wish to emphasize two points. First, it is important to appreciate that the uncertainties inherent to quantum theory do not arise because of gaps in our knowledge or understanding. It's not as though someday either the precision of our measurements or the level of our understanding will increase to the point where we can predict which photon will go this way, and which one will move another way. On the contrary, the more accurately we measure individual events, the clearer it becomes that the outcomes of those events are indeterminate.

Second, there is a pattern to these uncertainties, and these patterns account for the fact that the universe seems pretty orderly to us. Although we cannot predict the outcomes of individual events with any certainty, the overall outcomes of thousands of such events fall into very predictable statistical patterns. A mirror reflects ninety-five percent of the light hitting it, a subatomic particle is found within a certain distance of its atomic nucleus eighty-eight percent of the time, and at a particular energy level we can say precisely what the odds are that a chlorine atom will be in an excited, chemically reactive state. These probabilities give order to the physical and chemical world. They are the reasons why quantum indeterminacy does not produce universal chaos.

The apparent orderliness of nature does not mean that quantum events are irrelevant details dreamed up by physicists with too much time on their hands. Nor does it mean that quantum effects will always average out statistically, so that up here in the big world we inhabit, things will always be predictable and causal. Even though quantum effects are small by definition, we can easily design an apparatus that magnifies these subatomic happenings, transferring their inherent unpredictability to just about anything we like.

Figure 7.1 shows how easy it is to construct a machine built upon these events with actions that are inherently unpredictable. In the center of a chamber, we place a radioactive pellet that emits beta particles, which are energized, high-speed electrons. We choose the amount of the isotope carefully so that we can expect, on average, one radioactive disintegration per second for several weeks. Since each atom of the isotope is identical, we cannot tell which will be the next to send out a beta particle, and we also cannot predict the direction of the particle. If we surround the pellet with four separate detectors, each covering a quarter of the surface of the chamber, every time a beta particle hits one of the

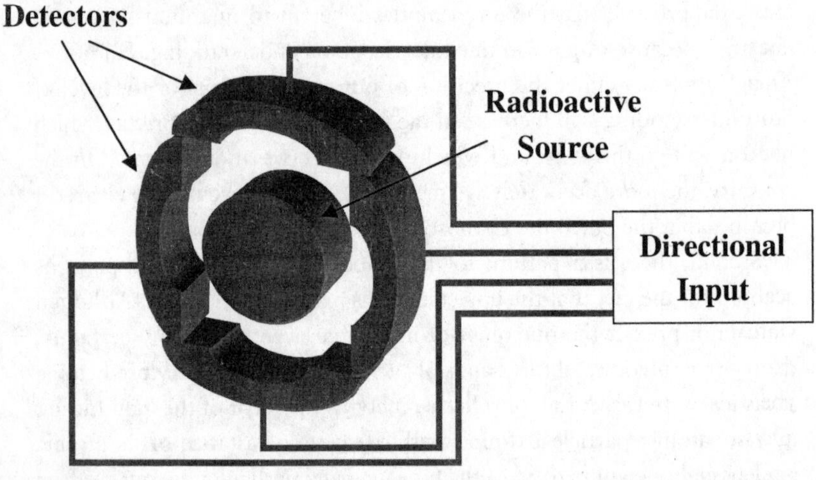

Figure 7.1. The nature of atomic events makes it easy to design a machine whose actions are unpredictable. A radioactive source, surrounded by four detectors, is attached to a computer system that interprets input from each of the detectors as a command to move in a different direction. Because the timing and the direction of radioactive disintegrations are unpredictable in principle, so are the movements of a device controlled by this system.

detectors it will send out a signal to a computing system that records the energized detector as a numerical value.

Next we write a simple program that controls the direction of movement in a remote-controlled mouse. The value sent by detector number one tells the mouse to move forward, number two backward, number three right, and number four left. As beta particles crash into the four detectors, we can now step back and watch our mouse execute a random walk across the experimental desktop.

The important thing about this demonstration is not that we can make a mouse move back and forth. Rather, it is that the movements of this large object are inherently and absolutely unpredictable. There is literally no way we can tell in advance what the moves will be, no matter how much we know about the apparatus. We have found a way to link the unpredictability of atomic-level events to observable everyday events that have meaning and significance in the "real" world. How much significance? That depends entirely on the mechanism. If we had liked, we could have linked the firing of a gun, the opening of a drawbridge, or any other macroscopic event to our four randomly activated detectors. What this contrived example shows is that any machine, even a natural one, that is influenced by quantum phenomena is inherently unpredictable. Quantum theory makes it possible to construct, according well-understood principles of physical law, a genuine machine whose actions are indeterminate.

Quantum theory has further implications, of course. Events large and small taking place in nature all the time are unpredictable, even in principle, since quantum phenomena are important in energy absorption, chemical bonding, and radioactive disintegration. The unpredictability of nature means that the clockwork assumptions of extreme materialism—namely, that knowing past history and mechanism relegates all events in the future to a predictable, rule-based playing out of chemistry and physics—are simply wrong. Wrong at their core, wrong in implication, and most especially wrong in terms of the philosophy that they support.

Quantum reality is strange, troublesome, and downright illogical, but its unexpected discovery solves one of the key philosophical problems faced by any religious person: How could a world governed by precise physical law escape a strictly deterministic future? Imagine for a moment that light really is just a continuous wave, and therefore every diffraction effect is predictable. Imagine that electrons occupy distinct, determinate positions around their nuclei. Imagine that radiometric decay is the result of an internal, analyzable, nuclear mechanism. If all these things were true, as they were presumed to be prior to 1900, they would render nature into a system of parts whose energies, positions, and velocities, if known, would be absolutely sufficient to predict each and every future position of the system. Reality would be set in stone.

Now, everybody realizes that the universe is way too big a place for us *actually* to determine the exact position, energy, and momentum of each and every particle. So there would always be a *practical* problem in making specific predictions for the future. However, practical problems don't bother philosophers—nor should they. If each and every future state of the universe is—*in principle*—predictable from the current state, then the role of any Creator was finished at the moment the universe came into being. There would be literally nothing for Him to do except to watch His clockwork spin. And there would be nothing for us to do, either, except blindly follow the determined instructions of the machines within us.

Fortunately, neither is the case. The indeterminate nature of quantum behavior means that the details of the future are not strictly determined by present reality. God's universe is not locked in to a determinate future, and neither are we. Sadly, few theologians appreciate the degree to which physics has rescued religion from the dangers of Newtonian predictability. I suspect that they do not know (at least not yet) who their true friends are!

What Is Life?

If quantum-induced uncertainty were a mere detail, something that showed up only in the rarefied environment of a physics laboratory, we might let it go as unimportant. But it's not. It underlies the behaviors of individual electrons and individual atoms, the basic stuff of matter itself. This being a book about biology, the sensible next question is whether or not the quantum revolution really has anything to do with life. It surely does.

Erwin Schrödinger, an Austrian physicist born in 1887, was a pioneer of quantum mechanics who fashioned one of the key mathematical tools with which physicists now describe reality. The Schrödinger wave equation makes it possible to calculate a wavelength for anything, even for things that we commonly think of as matter. For example, by solving the Schrödinger equation for an electron moving at one percent of the speed of light, we discover that the electron has a wavelength of 7 angstroms, several times larger than the diameter of a typical atom. This means that electrons will show wavelike properties such as diffraction and interference when they pass through matter and interact with particles on the atomic scale. By contrast, the wavelength of a compact car weighing 1,000 kilograms is more than a billion billion times smaller: 10^{-28} angstroms. Since the wavelength of a car is so much smaller than the dimensions of another car (about 3 meters, or $3 * 10^{10}$ angstroms), cars *do not* behave like waves when they interact. Instead, they act like particles—they crash.

In 1943, Schrödinger gave a series of lectures at Trinity College in Dublin, later published under the provocative title *What Is Life?* Correctly anticipating the direction that genetics and biochemistry might take in the next decade, Schrödinger's analysis was, above anything else, a direct attempt to argue that the events that take place within living organisms can be accounted for by physics and chemistry. He thought they could, and set about to suggest what sorts of evidence might be needed to show this was the case.

His starting point was to ask what he called an "odd, almost ludicrous question: Why are atoms so small?" No doubt every schoolchild, frustrated that matter is made of particles too small to be seen, has wondered the same thing. However, Schrödinger realized that the more important question was a little different: "Why must our bodies be so large compared with the atom?"[8] The answer he gave was that we are big relative to the atom to insulate us and our senses from atomic-level events.

Schrödinger reasoned that in order to work properly, organisms need to base their physiologies on predictable, physical laws. Imagine the chaos that would result if we could hear each atom that smashed against our eardrum, react to each water molecule touching our skin, see the very oxygen and nitrogen that we inhale with every breath.[9] Because we are so large relative to the atom, we can live in a world where our bodies only have to deal with events caused by hundreds of thousands of atoms. Because our parts are so large, we are able to live in an orderly, predictable world in which the random movements of molecules and the haphazard chaos of molecular movements disappear into the background. That's why atoms are so small.

But Schrödinger wasn't finished. If all of this was true, how could physics and chemistry account for the astonishing fidelity of biological inheritance? Even in 1943, biologists knew that genes were located on chromosomes, but they also knew—significantly—that genes themselves were very small. Schrödinger estimated that genes might have an upper limit of a million (10^6) atoms, an incredibly small number compared to the 600 billion billion (6×10^{20}) atoms in just a drop of water.

The tiny size of the gene posed two problems. First, how could an assembly of atoms so small be spared the molecular chaos that our own bodies manage to avoid only by being so big? How could a tiny gene find stability in a sea of crashing atoms? Second, if the atoms of the gene were bound into some sort of extra-stable array to solve the first problem, how could we then account for the tendency of genes to mutate, to change suddenly from one form to another?

Schrödinger's answers to both questions showed remarkable foresight. He argued that the hereditary substance would have to be an aperiodic crystal, a "code-script" not at all unlike the Morse code, in which individual clusters of atoms were arranged in a fixed sequence. To Schrödinger, the arrangement of this information-carrying aperiodic crystal was the key. It had to be unlike any other substance known to date. As he put it:

> Living matter, while not eluding the "laws of physics" as established up to date, is likely to involve "other laws of physics" hitherto unknown, which, however, once they have been revealed, will form just as integral a part of this science as the former.[10]

Today we can look back on this statement with a smile, knowing what Schrödinger did not, that the base-pairing mechanisms at the heart of the Watson-Crick model of DNA structure explain the stability of the genetic code-script brilliantly. These, if you like, are the "other laws of physics" he was hoping for.

If these other laws made the tiny code-script exceptionally stable, then how could mutations occur? Not knowing the exact chemical nature of the genetic material, Schrödinger couldn't say, but there was more than enough physical evidence to think about an answer.

> [It is] conceivable that an isomeric change of configuration in some part of our molecule, produced by a chance fluctuation of the vibrational energy, can actually be a sufficiently rare event to be interpreted as a spontaneous mutation. Thus we account, by the very principles of quantum mechanics, for the most amazing fact about mutations, the fact by which they first attracted de Vries's attention, namely, that they are "jumping" variations, no intermediate forms occurring.[11]

Today we know that Schrödinger was exactly right. The DNA molecule is structured in precisely a way that makes the behavior of individual atoms, even individual electrons, significant. When a mutation, a mistake in copying DNA, occurs, there are no intermediate chemical forms. One of the four DNA bases is changed completely into another base, and that change has a direct (and possibly permanent) effect on the code-script of the gene. As a result, events with quantum unpredictability, including cosmic ray movements, radioactive disintegration, and even molecular copying errors, exert direct influences on the sequences of bases in DNA.

This means that life is built around a chemistry that provides an amplifying mechanism for quantum events, not unlike our randomized mouse, but with far greater significance. Mutations, which provide the raw material of genetic variation, are just as unpredictable as a single photon passing through a diffraction slit. Schrödinger didn't wonder what this could tell us about evolution, but the lesson is clear—the fact that mutation and variation are inherently unpredictable means that the course of evolution is, too. In other words, evolutionary history can turn on a very, very small dime—the quantum state of a single subatomic particle.

Cheating Faust

A great story like the legend of Faust is destined to be told many times, in many ways. In additional to the classic originals, I have enjoyed the Faust theme retold as a caustic commentary on the legal profession in the film *The Devil's Advocate,* the desperation of a woeful baseball team in the play *Damn Yankees,* and the personal ambitions of a musician in Thomas Mann's novel *Doktor Faustus.*[12] In a few retellings of the Faust legend, such as Goethe's, the outcome is happy reconciliation;

in others, including the original, it is damnation. In some, Faust's quest is noble and enlightened; in others, he is driven by vanity and greed. In every case, the Faust legend remains at heart the story of a man willing to sell his soul to the devil for absolute knowledge.

I suspect that Faust captivates us repeatedly because we so easily see ourselves in his place. If Romeo and Juliet is all about the urge to reproduce, as at least one critic has said, then Faust concerns an urge almost as great—the desire to understand. In a world where all too few people care about science, it would be easy to understate the intensity of that passion. But visit any research laboratory, engage a young scientist in conversation about her work, and the emotional hunger for knowledge will come rising to the surface. There is no mistaking the brightness of that desire to understand.

More than one of my friends, baffled by a research problem, has told me they'd be willing to make the Faustian bargain. One friend has even imagined Mephistopheles laying out the terms of the deal: "You will achieve complete understanding of man and nature. You will possess this understanding for the rest of your life. And at the end of your life, you will repay me with . . . *your soul.*" My friend scratches his head, looks at old Mephistopheles, and asks, "So, what's the catch?"

Well, Mephistopholes notwithstanding, there really is a catch, and it has nothing to do with payback. You *cannot* make the bargain of Faust even if you want to. Quantum physics tells us that absolute knowledge, complete understanding, a total grasp of universal reality, will *never* be ours. Not only have our hopes been dashed for ultimate theoretical knowledge of the behavior of a single subatomic particle, but it turns out that in many respects life is organized in such a way that its behavior is inherently unpredictable, too. It's not just a pair of colliding electrons that defy prediction. The mutations and genetic interactions that drive evolution are also unpredictable, even in principle.

This is something biologists, almost universally, have not yet come to grips with. And its consequences are enormous. It certainly means that we should wonder more than we currently do about the saying that life is made of "mere" matter. The *true* materialism of life is bound up in a series of inherently unpredictable events that science, even in principle, can never master completely. Life surely is explicable in terms of the laws of physics and chemistry, just as Schrödinger hoped, but the catch is that those laws themselves deny us an ultimate knowledge of what causes what, and what will happen next.

This also means that absolute materialism, a view that control and predictability and ultimate explanation are possible, breaks down in a way that is biologically significant. It means that after we have obtained understanding of so much of the world around us, the ultimate mastery of even the tiniest bit of matter in the universe will always elude us. The fact that the key events of life itself pivot on these points of uncertainty makes it even more maddening.

I suppose we could look at it this way. Strong, absolute, God-excluding materialism offers us a bargain in simple terms. We should regard science as a way of knowing that excludes theism, because science can explain how things work and theology cannot. Where theology promises only miraculous explanations for nature, science can show you the *real* explanations. The refraction of white light is caused by moisture droplets in the air, and from such calculable physics comes the beauty of the rainbow. Exchange your belief in God for a material theology of disbelief, and complete knowledge will be yours. Like the idea so far? Faust certainly would have. But Faust might have been skeptical enough to ask if the offer was genuine, if the knowledge to be obtained at the price of his soul would indeed be complete.

Imagine his reaction if Mephistopheles were to own up to the fact that the best he could offer was a statistical description of events, a calculation of the probabilities that emerge only when hundreds or thousands of unpredictable quantum events are averaged out. How complete can knowledge be when it cannot link cause and effect for something so basic as the movement of an electron? How can it claim to fully describe the universe when it cannot tell me whether something as humble as a single atom of carbon will persist into the next instant or disintegrate in a burst of radioactive energy? For any sensible Faust, the promises extended by this Mephistopheles are bogus. The core assumptions supporting the "scientific" disbelief of the absolute materialist are wrong, even by the terms of science itself.

The Mind of God

What does all of this mean for those of us who see science as the paramount way of knowing and understanding nature? Clearly it means that science's ultimate goal, complete knowledge, will never—indeed, *can never*—be realized for even the smallest of nature's individual parts. That's a given, and it matters to us today because the classic science-versus-religion mentality we still encounter took shape within a nineteenth century mind-set that saw any advance in scientific understanding as a threat *per se* to the idea of God. At one time, that might have made sense. Had Darwinism prevailed in a strictly Newtonian world, we would have been left to view our biological history—and our future—as strictly determined as the outcome of any shot on a billiard table.

Well, Darwinism *has* prevailed, and has done so dramatically. And by reflex, the nineteenth century logic that its triumph would be unavoidably tied to the downfall of religion has kicked in, too. As one oft-told tale goes, a member of British society, when told about Darwin's theory, gasped at its implications: "Let us hope that it is not true. But if it is, let us hope that it does not become generally known!"

This intellectual reaction is common because the science-versus-religion argument has been framed within the classical view of physics that shaped such dialogues in the last century. Such misunderstandings persist because the field of biology has not yet fully reacted to the revolution in modern physics. It hasn't had to, because nearly all experimental biology takes place in a realm where quantum effects are averaged out into statistical laws.

A critic might say this is precisely because the indeterminacy of modern physics doesn't really matter in biology. But that's not true at all. It's just that it crops up in ways we generally fail to recognize. *Wonderful Life,* Stephen Jay Gould's book on the Burgess shale, provides us with a perfect example.

The Burgess shale is a remarkable formation of sedimentary rock in British Columbia that contains an exquisite set of animal fossils from the Cambrian period of geologic history, roughly 530 million years ago. Gould described a groundbreaking reinterpretation of the fossils in the late 1980s by a number of paleontologists, including Harry Whittington, Derek Briggs, and Simon Conway Morris. Their elegant analytical work on Burgess shale specimens showed that many of the fossils once thought to be members of existing animal phyla were different, very different, from anything alive today. Research on the Burgess specimens continues, and recent finds of Cambrian fossils in China have increased our understanding of Cambrian biodiversity.

This is more than just a bookkeeping change. The *phylum* is the next-to-highest level of biological classification. Only *kingdom* is higher, and all animals belong to the same kingdom (Animalia). Each phylum represents a distinctive body plan. The recognition that this little grouping from the floor of an ancient sea contained animals from as many as nine completely extinct phyla was a surprise that could be savored only by specialists—until Gould used the shale to ask an interesting question. When we look at these novel body plans, when we compare the specimens from extinct phyla with the ones that persist to this very day, can we understand why some succeeded and some perished? Gould's answer is no, we cannot.

According to Gould, this is not just because of practical difficulties in identifying which forces were most powerful in shaping the course of life 600 million years ago. There's more to it than that. It's because even if we could reconstruct exactly what that ancient world was like, even if we could know everything about the relative fitness, differential reproduction, and physiologic efficiency of these ancient animals, we still couldn't predict which would have been the winners and which the losers in the evolutionary sweepstakes. We couldn't have known because the results of that ancient lottery were, to some extent, a matter of chance. And, according to Gould, that's how evolution works—in unforseeable ways.

> I believe that the reconstructed Burgess fauna, interpreted by the theme of replaying life's tape, offers powerful support for this different view of life: any replay of the tape would lead evolution down a pathway radically different from the road actually taken.[13]

Gould's metaphor of the tape of life, in his view, "promotes a radical view of evolutionary pathways and predictability."[14] He obviously believes that he has added something new and radical to our understanding of evolutionary history. What he has actually done, of course, is to make a more modest observation on the contingent nature of history in general. Every event in time, no matter how small, exerts some influence on the future course of history. If some of those events are inherently unpredictable, then the course of history is going to be just as unpredictable.

The observation that Gould finds so remarkable follows naturally from the cause-and-effect links that extend upward from quantum physics through chemistry and biochemistry into the undirected input of variation into living, genetic systems. Gould may not have recognized the physical roots of his observations, but they are nonetheless there for all to see. The natural history of evolution is unrepeatable because the nature of matter made it unpredictable in the first place. Wind that tape back, and it will surely come out differently next time around, not just for the Burgess shale, but for every important event in the evolutionary history of life.

Does this tell us something important about reality? Does the pervasiveness of quantum indeterminacy enable us, in a sense, to "know the Mind of God," as Stephen Hawking so memorably put it? In some respects, I think it does. Albert Einstein's oft-quoted comment, "God does not play dice," reflected his profound discomfort with an indeterminate world. Einstein believed as a matter of faith that physics did have something important to tell us about ultimate reality, and didn't like the idea that *any* event was truly indeterminate. In the end, Einstein's dissent failed. His distaste for chance notwithstanding, the physical evidence for the indeterminate nature of quantum reality won out. As Neils Bohr said of his friend's remark, "Who is Einstein to tell God what to do?"

I suspect that the seeming randomness derived from quantum reality is one of the things that makes evolution so offensive to its frustrated opponents. They have a visceral dislike of any scenario in which the present is a random product of the past, of a history in which literally anything could have evolved, so that we just happen to be here. If we're just the product of chance, their reasoning must go, then how could we be the intentional creations of a loving God who *meant* us to be here? This concern, clearly, is at the very heart of creationist objections to evolution.

Part of their problem surely comes from the act of confusing "random" with "indeterminate." In ordinary speech, when we say that something is random, we generally mean that *anything* can happen, with all outcomes having equal probabilities. The shuffling of a deck of cards, in the hands of a competent and honest dealer, produces a truly random outcome in which any of the many (fifty-two-factorial) arrangements of the shuffled deck are equally probable. Physical events, generally speaking, are not at all random in this sense because all outcomes are *not* equally probable. The possible locations of an electron around an atomic nucleus are given by a probability wave that tells us which locations for the electron are most likely and which are so unlikely that they need not be considered as possible. Until we actually observe the location of the electron, we cannot tell where it will show up in that probability wave, and that is why its actual location is indeterminate. But this does not mean that an electron could, with equal probabilities, be just *anywhere*. Events at the atomic level are indeterminate, but not random—they follow understandable statistical patterns, and those patterns are the ones we sometimes elevate to the status of physical laws.

Remarkably, what the critics of evolution consistently fail to see is that the very indeterminacy they misconstrue as randomness has to be, by any definition, a key feature of the mind of God. Remember, there is one (and only one) alternative to unpredictability—and that alternative is a strict, predictable determinism. The only alternative to what they describe as randomness would be a nonrandom universe of clockwork mechanisms that would also rule out active intervention by any supreme Deity. Caught between these two alternatives, they fail to see that the one more consistent with their religious beliefs is actually the mainstream scientific view linking evolution with the quantum reality of the physical sciences.

We need not ask if the nature of quantum physics *proves* the existence of a Supreme Being, which it certainly does not. Quantum physics does allow for it in an interesting way, and certainly excludes the possibility that we will ever gain a complete understanding of the details of nature. We have progressed so much in self-awareness and understanding that we now know there is a boundary around our ability to grasp reality. And we cannot say why it is there. But that does not make the boundary any less real, or any less consistent with the idea that it was the necessary handiwork of a Creator who fashioned it to allow us the freedom and independence necessary to make our acceptance or rejection of His love a genuinely free choice.

Committed atheists like Richard Dawkins would attack with ridicule any suggestion that room for the work of a Deity can be found in the physical nature of reality. But Dawkins's personal skepticism no more disproves the existence of God than the creationists' incredulity is an argument against evolution. What matters is the straightforward, factual, strictly scientific recognition that matter in the universe behaves in such a way that we can *never* achieve complete knowledge of any fragment of it, and that life itself is structured in a way that allows biological history to pivot directly on these tiny uncertainties. That ought to allow even the most critical scientist to admit that the breaks in causality at the atomic level make it fundamentally *impossible* to exclude the idea that what we have really caught a glimpse of might indeed reflect the mind of God.

Deism, Real and Imagined

Ironically, the anti-evolutionists, just like the extreme materialists, seem to work in an intellectual world uninformed by modern physics. To their way of thinking, scientific materialism necessarily implies a complete, causal, mechanistic description of natural reality. They have seen the profound successes of biology in providing material explanations for the nature of life, and they are worried. Molecular biology has succeeded in finding a material mechanism for the details of inheritance, and is now in the process of discovering the rules by which DNA-coded instructions are executed in development—providing ever more detailed glimpses into the drama by which a single cell grows into an adult organism.

In 1995, one of my colleagues ended a seminar on developmental biology by summarizing his own recent work but then mocking its limited scope, quipping that "for the really big things, like telling right from left, all we can say is that the embryo still relies on God." What he meant was that for all our understanding of embryonic development, some basic questions, like how a few cells lay down an axis along which the left and right sides of the body differentiate, remain unanswered. We might as well just say that God did it. Like everyone else in attendance, I appreciated his self-deprecating modesty. Only three years later, I had occasion to remember his words with a chuckle. The mechanism by which the vertebrate embryo tells right from left was finally discovered in 1998.[15] I suppose you could say that God had lost another job, but for anyone who had seriously followed the progress of the life sciences, this small advance in our understanding was just another page in a long story. Any idea that life requires an inexplicable vital essence, a spirit, an *elan vital,* has long since vanished from our lives and laboratories, a casualty of genetics and biochemistry.

This clearly has been noticed by the opponents of evolution, who see each new materialist, scientific explanation for a natural phenomenon as a retreat for the primacy of God. They are willing to fix on anything that seems unlikely to have a material explanation as proof of the divine agency they seek. And given biology's remarkable successes in accounting for life's workings *today*, they are obligated to find those miracles in the past. In this respect, they are the unlikely and unwitting allies of vigorous atheists like Dawkins, Provine, and Lewontin who would agree enthusiastically that a complete success for materialist science condemns any search for God to ultimate failure.

One hundred years ago, the very same logic might have pointed to the sun's fire as evidence of the divine—and in the days before fusion, who could have argued against it? Wouldn't an open-minded science have admitted that the sun's fire, inexplicable by any analogy to earthly chemistry, might indeed be a miracle? Well, sure . . . except, of course, for anyone who had watched the steady progress of science in explaining physical phenomena, and therefore might have suspected that it could be counted upon to extend its reach to our nearest star. Which is exactly what it did.

The anti-evolutionists have, in essence, treated the origins of species as a kind of unexplained solar fire in which the inability of science to provide immediate explanation becomes proof for the existence of God. God, therefore, becomes the default explanation, their refuge of choice, for any event in natural history we cannot yet explain. The successes of scientific materialism, especially in biology, have led the anti-evolutionists to propose that the only feasible way for the Creator to have interacted with living organisms was to have *made* them, sometime in the distant and unobservable past. Hence their insistence that the origins of species, the appearance of the first cells, and even the biochemical design of the flagellum must be the direct and miraculous work of a designer.

I have done my best to hammer this desperate strategy with facts in the first few chapters of this book. Strong scientific evidence exists to show that evolution can account for the natural history of life. No doubt we have a great deal yet to learn, but we really do understand the essence of the sun's fire, and we really do understand the forces that brought our species into existence. There is no point in pretending that we don't.

Nonetheless, to creationists like Johnson, Behe, and Dembski, a fixation with biological origins is an essential element of their theology, because they argue that otherwise the role of God could be nothing more than a super-architect, a great designer who fashioned a self-contained universe, gave it a shove, and then let it run. They ridicule the notion of a passive, deist God whose activity comes to an end once the essential constants of the universe are fixed. In Dembski's words:

> The world is in God's hand and never leaves his hand. Christians are not deists. God is not an absentee landlord.[16]

Richard Dawkins agrees. Reacting to those who might take satisfaction in a view of God as responsible merely for the physical constants of the universe, he writes:

> The trouble is that God in this sophisticated, physicist's sense bears no resemblance to the God of the Bible or any other religion. If a physicist says God is another name for Planck's constant, or God is a superstring, we should take it as a picturesque metaphorical way of saying that the nature of superstrings or the value of Planck's constant is a profound mystery. It has obviously not the smallest connection with a being capable of forgiving sins, a being who might listen to prayers, who cares about whether or not the Sabbath begins at 5 pm or 6 pm, whether you wear a veil or have a bit of arm showing; and no connection whatever with a being capable of imposing a death penalty on His son to expiate the sins of the world before and after he was born.[17]

Just like the creationists, Dawkins rejects any notion that a deist God could be compatible with the essential tenets of Western religion. For what it's worth, so do I. But something interesting is going on here. At the heart of their anti-evolutionism, the creationists have hidden a stunning inconsistency in their own logic.

Consider this: Creationists would reject any notion that God is unable to act in the world today. Indeed, Christianity, like Islam and traditional Judaism, regards the continued, personal activity of God to be an essential element of belief. Now, let's step back a bit and think about this.

As a matter of unshakable faith, they believe that God can act in the world at the present time. And that, presumably, He can work His will in any way He likes—with power or with subtlety, by works of nature, or by the individual actions of His creatures. The very same people, bowing to the explanatory power of molecular biology and biochemistry, would also agree that life today can be understood as a wholly material phenomenon.

None that I know of would reject the proposition that a single fertilized egg cell—the classic specimen of developmental biology—contains the full and complete set of instructions to transform itself into a complex multicellular organism. Neither would any respectable creationist challenge the assertion that every step of that developmental process is ultimately explicable in terms of the material processes of chemistry and physics. Miracles aren't required—the complexities of molecular biology will do just fine.

This means that the biological world of today, which we can test and study, analyze and dissect, is one that works according to purely material rules. But this world is also one in which believers, as a matter of faith, accept sincerely the tenet that God can and does work His will. Obviously, they do not see any conflict in the idea that God can carry out the work He chooses to in a way that is consistent with the fully materialist view of biology that emerges from contemporary biology. Neither do I.

But there's the rub. Curiously, they somehow regard those very same mechanisms—adequate to explain God's power in the present—as inadequate to explain His agency in the past. For some reason, God acted in the past in ways that He does not act in the present, despite the fact that we assume in the present that he can do *anything* and *everything*. This inconsistent reasoning is at the heart of their desperation to show that evolution—which depends on the material mechanisms of biochemistry and genetics—could not have created the multitude of new species that have appeared throughout the geologic record of life.

Such reasoning shows a curious lack of faith in the creative power of God. Creationists act as though compelled to go into the past for evidence of God's work, yet ridicule the deistic notion of a designer-God who's been snoozing ever since His great work was finished. They want a God who is active, and active now. So does any believer. But why then are they so determined to fix in the past, in the supposed impossibility of material mechanisms to originate species, the only definitive signs of God's work? If they believe in an active and present God, a God who can work His will in the *present* in ways consistent with scientific materialism, then why couldn't that same God have worked His will in exactly those ways in the *past*?

The more sensible and self-consistent position, scientifically *and* theologically, would be quite different from theirs. The real, actual, working world that we see around us is one that is ruled by chemistry and physics. Life works according to its laws. If God is real, *this* is the world He has to work in. Therefore any effort to view God's work in light of modern science must find a way to understand how His will can be accommodated at all times, not just in some distant past.

Sadly, the creationists reject this notion, which leads to a remarkable conclusion. Namely, that the true deists, the ones who presuppose a period of creative power only in the past followed by passive inaction in the present, are actually the creationists. By holding steadfastly to the view that the observable, material processes of evolution could never explain the appearance of new species, they cast themselves, ironically, into the deist mold. Their view of God places unintentional but profound limits upon His actions in the present. They require that His power to originate species be banished to a period in the past, and in so doing limit His ability to act in the present. Their God created and destroyed countless species in bursts of miraculous, nonmaterial activity in the distant past, but in their deist present, He is curiously constrained from such actions.

Neither Christian theology nor scientific theory contains justification for such limitations. In the traditional view of God's power held by all Western religions, God's presence and His power are part of the continuing truth of existence. What this means, in plain and simple terms, is that ordinary processes, rooted in the genuine materialism of science, ought to be sufficient to allow for God's work—yesterday, today, and tomorrow.

A Science of Belief

At the end of the day, the stunning successes of the life and material sciences have forever changed our picture of the world. We have grown up as a species, if only because we now know where and how to look for answers. We therefore seek materialist and scientific explanations for the things around us. The superstitions that hobbled our ancestors have been overcome, and our successes have given us stewardship of this planet to a degree that biblical writers could hardly have imagined.

With the smoke of so many victories in the air, why not celebrate the triumph of materialism? Why not proclaim the absolute death of religious belief? Why not assert, in Richard Lewontin's words, that it is time to accept science "as the only begetter of truth?" Why not take Faust's bargain? Part of the reason, as we have seen, is that science has an inherent limitation in how much it can tell us about reality. Because of this, it would be a serious scientific mistake to assert that the successes of molecular biology and biochemistry in explaining how living things work, and the complementary successes of evolution in explaining how they got that way, can justify a triumphantly anti-theistic materialism. No matter how hard we probe, the peculiar quantum nature of reality does not allow us to predict the behavior of even a single electron with certainty. And this uncertainty, which theory suggests we can never overcome, prevents science from ever attaining a complete understanding of nature.

"So what?" the extreme materialist might say. Why does it matter if a few events at the atomic level are unpredictable? Or even if life itself is built upon an apparatus that amplifies those events into significance? It matters because materialist science, even in principle, cannot tell us *why* the universe of matter is structured in a way that prevents us from understanding it fully. Or why nature forever entangles the observer with the system he seeks to understand. Or even why we should concern ourselves with seeking the answers to such questions.

We began this chapter by asking whether the advances of science, which have so thoroughly displaced the pagan superstitions of animism, have now ruled out even a Western, monotheistic God. Had classical physics reigned triumphant, they might well have. Unexpectedly, the ultimate physics of nature did not complete a chain of cause and effect. It left an open window on events, a break in causality that is significant not because science cannot master a few tiny details of the physical universe—but because it cannot even address the question of why nature should be constructed along such elusive lines. In the final analysis, absolute materialism does not triumph because it cannot fully explain the nature of reality.

It would be foolish to pretend that any of this rigorously proves the existence of God. If it did, we should expect missionaries to win souls by explaining two-slit diffraction and by showing the derivation of Planck's constant. But the tools of science itself have discovered that scientific materialism has a curious, inherent limitation. And we are certainly left to wonder what to make of that. It could be just a puzzling, curious fact about the nature of the universe. Or it could be the clue that allows us to bind everything, including evolution, into a worldview in which science and religion are partners, not rivals, in extending human understanding a step beyond the bounds of mere materialism.

8

THE ROAD BACK HOME

Ironically, when I have publicly advanced the idea that God is compatible with evolution, I find that my agnostic and atheistic colleagues are generally comfortable with such ideas, but many believers are dumbfounded. "How can you reconcile divine will with a random, chance process like evolution?" is a common question. Biologists, who appreciate the importance of historical contingency in evolution, might well agree: "Sure, God could have produced us by evolution—so long as He was willing to take His chances that He might have wound up with nothing better than big-brained dinosaurs."

The believers have detected something important. From time to time, the National Academy of Sciences spins out high-minded reports on the inclusive compatability of religion and science. I would like to think that these documents reflect a genuine appreciation for the value of religious thought, but those who know firsthand the sentiments of many evolutionists (Dawkins: "I am against religion because it teaches us to be satisfied with not understanding the world.") may be forgiven if they take those fine words as a cloak for something else.

On an emotional level, neither the formulaic acceptance of religion by the National Academy nor an admission like E. O Wilson's that science can never *completely* rule out God sound truly convincing.

A scientifically compatible version of religion could mean something weak and pale and very different from the forms of religion embraced by most Americans. Indeed, the variety of religious beliefs is so great that one might fairly wonder if those scientists who proclaim compatibility with religion have something very nonreligious in mind. When I tell my students I believe in God, they suppose that I couldn't possibly mean anything traditional, but rather something smart, modern, and sophisticated. Something subtle. Maybe I mean that God is love, or God is the universe itself, or, being a scientist, maybe I mean that God is the laws of nature.

Well, I don't. Such views, however carefully stated, dilute religion to the point of meaninglessness. As Carl Sagan noted, such a God would be "emotionally unsatisfying . . . it does not make much sense to pray to the law of gravity."[1] Nor, I might add, does it make any sense to pray to the second law of thermodynamics, which has never given me a break, and probably never will.

Such "Gods" aren't God at all—they are just clever and disingenuous restatements of empirical science contrived to wrap an appearance of religion around them, and they have neither religious nor scientific significance. In this book I am interested in a traditional view of God—the one described by the great Western monotheistic religions. That's the God that believers wonder about, that's the one they pray to, and that's the one who *seems* to be threatened by evolution. And that is the God whose actions, intentions, and existence I wish to consider in this chapter.

He Who Am

Many of my scientific colleagues would think me crazy to write *anything* about God. Too many dangers, too many traps. They tell me you can't say *anything* even remotely scientific about religious issues like the virgin birth, the trinity, the role of the prophets, or the meaning of scripture. And even if you could, anything you say would get people upset—lots of people. Point taken.

The world has many religions but just one science, and that tells us something about both. I do not mean, of course, that all scientists agree all the time on all issues. Science thrives on controversy and disagreement. But there is one science, worldwide, in the sense that scientific issues can be addressed by a common set of principles. All scientists use experiment, observation, and analysis to settle scientific issues, and ultimately all subscribe to a final test of their theories in the reality of nature. Among scientists, whatever our individual prejudices and however differently we might see things, we can get together and talk about genuine scientific issues, like the control of DNA replication, using a common language of understanding. However imperfect, there really is a single, worldwide, scientific culture.

This isn't even remotely true for religion. In fact, many skeptics point to the extreme plasticity of human faith as a prime reason for their own disbelief in God. Could God be real when equally sincere people of faith hold different opinions on issues so basic as to who might be God's chosen people, whether the Messiah has come, or whether it is sinful to make a graven image? Given so many different versions of God, so many religions, so many churches, how can anyone hope to talk sensibly of the relationship between God and science?

If our goal is to analyze in fine detail the precise fit between evolutionary biology and each and every shade of religious belief, we might as well give up now. But let's not. It seems to me that, whatever their doctrinal differences, the three great Western religions share a core of belief. With those common elements in mind, we can approach the issues that matter most.

The first common belief is the primacy of God in the universe. Judaism, Christianity, and Islam all believe in a genuine, personal God who created the universe and everything within it by an act of His own volition. Their God is eternal, but the world He created has a distinctive beginning and may eventually come to an end.

Second, we exist as the direct result of God's will. We are the intentional creations of God, creatures with physical bodies but immortal, spiritual souls. He has endowed us with free will, allowing us to choose good or evil, to love God or reject Him.

Third, God has revealed Himself to us. God has spoken through the prophets, or by the words of Jesus, or through Muhammad, and by direct, personal contact with individuals of faith. In each age He finds a way to bring His message directly to us.

Each of these religions also makes a crucial distinction between the spiritual and the material. As Saint Thomas Aquinas pointed out, the very idea of God requires that He be a nonmaterial being; in other words, a spirit. This also means, for any religious believer, that there is a spiritual reality that surpasses the physical reality of nature.

Some religions have explicit ways of showing this. For example, many Christian faiths carry out a ritual in which they believe ordinary bread and wine are changed into the body and blood of Jesus. They do not mean, to speak like a biologist, that the consecrated wine contains hemoglobin, red cells, or platelets. They do not mean that the gluten and plant protein of the transubstianted host is suddenly changed into the collagen and fibrin of human tissue. At the same time, they do not mean that the transformation is merely symbolic. The bread of a communion service, they believe, actually becomes the spiritual flesh of Christ. They believe, as an element of faith, that the spiritual is just as real as the material.

Finally, each of these great faiths believes that God is active in the world in a personal sense. Their God is one who answers prayers—or, keeping a Garth Brooks song in mind, sometimes chooses *not* to answer them. Their God provides hope and inspiration, performs miracles, and provides the resources needed for individual salvation. Let there be no doubt about it: The God I am speaking of is the one who whispered to Moses from the burning bush, "I am He whom am." Most readers of this book, whether they believe in Him or not, will recognize Him immediately.

The question is, what does He think of evolution?

In the Beginning

Prior to the age of science it might have been possible to assert that the universe, if not the earth itself, had always existed. The recurrence of seasons, the changeless patterns of stars, the ebb and flow of cycles in nature all suggested stability, eternal and pervasive. Time itself, the most fundamental of all variables, might have seemed to be infinite, flowing seamlessly from an endless past into an eternal future. Then up popped a little problem from thermodynamics. The enormous energy expended by the sun and its multitude of sister stars had to come from somewhere, for eventually it would burn down to nothing. The Newtonian universe might have been a clockwork, but in energetic terms at least, the spring of that clock must somehow have been wound up. Such a job would have required even the passive God of deism to roll up His sleeves and get busy—at least once.

The vague notion of an energetic beginning became much more specific with the discovery of radioactivity. The persistence of radioactive disintegration deep within the earth suddenly explained why our planet's core refused to cool—but it also placed an upper limit on the planet's age. To take just one example, when geophysicists determined that the half-life of ^{235}U, a radioactive isotope of uranium, was roughly 700 million years,[2] it became self-evident that the earth could not be infinitely old. Only one-half of the planet's original ^{235}U would remain if the planet were 700 million years old. One-quarter if its age were 1.4 billion. One-sixteenth if the earth had been formed 2.8 billion years ago. In fact, if the planet were older than 10 billion years, the amount of ^{235}U should have dwindled to the point of undetectability. Well, it hasn't. There is indeed very little of the stuff around, but still more than enough to fashion the implements of nuclear power and nuclear warfare, as our modern history records. The earth had a beginning, it was "only" a few billion years ago, and the date of that beginning is written like a signature into the very ground upon which we stand.

In 1929, Edwin Hubble extended the notion of a beginning to the universe itself. He examined light reaching the earth from distant galaxies, and found that the Frauenhofer lines of such light are shifted, sometimes substantially, into the red region of the spectrum. The most sensible explanation for this shift, as Hubble realized, is that the distant galaxies are moving away from the earth at great speeds. The greater the red shift, the faster the galaxy is moving, just like a speeding locomotive whose whistle deepens in pitch as it moves away from an observer. To Hubble's initial amazement, the most distant galaxies were moving the fastest—almost as though everything in the universe had been blown apart from a great initial explosion at a single point in space and time, an explosion known popularly as the big bang.

So, what caused the big bang? What set the fuse? What existed prior to that great explosion? Good questions, and questions for which an astonishingly large number of people would like to know the answers. Stephen Hawking's popular book *A Brief History of Time*[3] was an ambitious attempt to answer these and other questions on the initial state of the universe. Its technical density notwithstanding, it became a best-seller, which is unusual for any scientific book and demonstrates the intensity of public interest in such subjects. Hawking's successful work was followed by a number of more readable popular accounts of cosmological theory, including Timothy Ferris's playfully titled *The Whole Shebang*.[4]

Although these accounts are laced with heavy doses of speculation, they also involve a tantalizing amalgam of serious science—astronomy, cosmic history, and basic physics. The reason, of course, is that when one makes a run backwards in time to the moment just before the big bang, one must imagine inconceivable amounts of mass and energy concentrated at a single point in space. To understand the behavior of matter and energy under such conditions requires that physical principles—developed in the relaxed, scattered, low-energy environments of the universe today—be stretched right up to (or maybe just beyond) the breaking point. Trying to get an experimental handle on how matter behaves under such conditions has been, in fact, one of the principal reasons for trying to build larger and larger accelerators to smash particles into each other.

One of the most remarkable findings of cosmological science is that the universe did have a beginning, and a spectacular beginning at that. Discussions of first causes used to be dry philosophical constructs, theoretical arguments against an infinite regression of events backwards in time. The big bang made the first cause real. It placed a wall at the beginning of time, closing to inquiry (but not, of course, to speculation) all events that might have occurred before that cosmic explosion. In the view of many scientists, the big bang casts a distinctly theological light on the origin of the universe. Here's how astrophysicist Robert Jastrow put it:

> At this moment it seems as though science will never be able to raise the curtain on the mystery of creation. For the scientist who has lived by his faith in the power of reason, the story ends like a bad dream. He has scaled the mountains of ignorance; he is about to conquer the highest peak; as he pulls himself over the final rock, he is greeted by a band of theologians who have been sitting there for centuries.[5]

Jastrow's theologians, of course, were sitting there telling the scientists, "We told you so!" What was really happening was more complex.

Either there is a God, and the big bang dates the moment of His creation of the universe, or there is a tendency of matter to create itself from nothingness. If that is the case, the big bang merely marks the moment of that self-creation or the latest oscillation in a grand series of cosmic cycles: big bang followed by big crunch, followed by yet another big bang.

If cosmology provided us with a way to distinguish between these two extreme alternatives, we might then wait for the scientific word from on high on the status of the Almighty. Unfortunately, it doesn't, and we can't. As Timothy Ferris puts it:

> So it seems reasonable to ask what cosmology, now that it is a science, can tell us about God.
>
> Sadly, but in all earnestness, I must report that the answer as I see it is: nothing. Cosmology presents us neither the face of God nor the handwriting of God nor such thoughts as may occupy the mind of God. This does not mean that God does not exist, or that he did not create the universe, or universes. It means that cosmology offers no resolution to such questions.[6]

Cosmology, despite the best hopes of believers, will not resolve the issue of God's existence, or for that matter, of the meaning and purpose of life. You may share Ferris's disappointment in discovering this. But I see the situation differently. What we do know is that the universe, our planet, and life itself all had distinct origins, their own beginnings, and so did we. This does not prove that the ultimate beginning must be attributed to the divine, but neither does it provide any scientific basis for ruling God out. Remember that first common belief of Western religion, the primacy of God in bringing the universe into existence? Whatever its faults and uncertainties, science has confirmed, in remarkable detail, the distinctive beginning that theology has always required. Maybe Jastrow's theologians have reason to smile, after all.

STACKING THE DECK?

Emotionally, one of the most powerful arguments against the existence of God would seem to be the structure of the universe itself. Here we sit, having once thought ourselves the center of that universe, only to learn that we occupy a tiny planet swinging around a star of below-average intensity, at the periphery of a nondescript galaxy dwarfed by thousands of others larger and more magnificent than our own. If there is a God, and if He created this universe just for us, He seems to have waited billions of years to get around to us, and when He did, He stuck us off in an insignificant cosmic backwater.

The scale of the universe might argue that we are insignificant, but the physical structure of the very same universe seems to say something quite different. Gravity is a good example. Newton showed that the force of gravity between two bodies is related to the product of their masses, divided by the square of the distance between them. Specifically:

$$F = \frac{m_1 * m_2}{d^2} * g$$

This simple equation has a good deal of common sense built into it. For example, if I have a mass of 80 kilograms and another person weighs just 40 kilograms, gravity will pull on me twice as hard as it pulls on him. If we travel to a planet one-fourth the mass of the earth, gravity there will be about four times weaker than it is on earth. And as we move away from the earth, the force of gravity will diminish by the square of the distance. At a distance of 2 million kilometers, the force of the earth's gravity is only one-fourth of its strength at 1 million kilometers. This is all well and good. At a glance this equation tells us all of these things about the *relative* strength of gravity. But what about its absolute strength? To find this, we have to multiply the masses and positions of the bodies by the universal value of *g*, the gravitational constant, which determines the ultimate strength of gravity everywhere in the universe. The value of *g*, in case you are interested, is $6.67 * 10^{-11}$ (the units of *g* are: Newtons – Meters squared over kilograms).

So much for remedial physics—now let's have some fun. Does *g* *have* to be $6.67 * 10^{-11}$? What if *g* were a little larger or a little smaller? It turns out that the consequences of even very small changes in the gravitational constant would be profound. If the constant were even slightly larger, it would have increased the force of gravity just enough to slow expansion after the big bang. And, according to Hawking, "If the rate of expansion one second after the big bang had been smaller by even one part in a hundred thousand million million it would have recollapsed before it reached its present size."[7] Conversely, if *g* were smaller, the dust from the big bang would just have continued to expand, never coalescing into galaxies, stars, planets, or us.

The value of the gravitational constant is *just right* for the existence of life. A little bigger, and the universe would have collapsed before we could evolve; a little smaller, and the planet upon which we stand would never have formed. The gravitational constant has just the right value to permit the evolution of life.

Our luck didn't stop there. Gravity is one of four fundamental forces in the universe. If the strong nuclear force were just a little weaker, no elements other than hydrogen would have been formed following the big bang. If it were just a little stronger, all of the hydrogen in the universe would be gone by now, converted into helium and heavier elements. Without hydrogen, no sun, no stars, no water.

If another fundamental force, electromagnetism, were just a little stronger, electrons would be so tightly bound to atoms that the formation of chemical compounds would be impossible. A little weaker, and atoms would disintegrate at room temperature. If the resonance level of electrons in the carbon atom were just four percent lower, carbon atoms themselves would never have formed in the interiors of stars. No carbon, no life as we understand it.

As astronomers B. J. Carr and M. J. Rees put it in a 1979 review in the scientific journal *Nature*:

> The possibility of life as we know it evolving in the Universe depends on the values of a few basic physical constants—and is in some respects remarkably sensitive to their numerical values.[8]

It almost seems, not to put too fine an edge on it, that the details of the physical universe have been chosen in such a way as to make life possible. Recognition of this has led to the formulation of what is known as the "anthropic principle." The physical constants of the universe in which we live *have to be* favorable to human life, because if they were not, nobody would be around to observe them. In other words, the very fact that we are here to make a fuss means that the physical constants of the universe were set up in a way that made our existence possible.

Now, before we get carried away, let's keep in mind that the physical constants we have been discussing are just that—constants. We have no way of knowing for sure how those constants were determined, whether or not they might be different in another universe, whether they were fixed by the conditions of the big bang itself, or whether they reflect an unchanging physical reality that predated the big bang and the origin of our universe. We also have to keep our minds open to the possibility that future advances in physics may one day explain the apparent coincidences that seem to link many of these constants.

Having considered all that, Carr and Rees are still amazed.

> Even if all apparently anthropic coincidences could be explained in this way, it would still be remarkable that the relationships dictated by physical theory happened to be those propitious for life.[9]

Physicist Freeman Dyson is dazzled as well.

> I do not feel like an alien in this universe. The more I examine the universe and study the details of its architecture, the more evidence I find that the universe in some sense must have known we were coming.[10]

In a way, the anthropic principle plucks us out of that cosmic backwater and once again makes us the center of the universe. No less an authority than Stephen Hawking has said:

> The odds against a universe like ours emerging out of something like the Big Bang are enormous. I think there are clearly religious implications.[11]

Naturally, the notion that we might be here as the result of an intentional *choice* of the constants that govern physical laws is a little too much for nonbelievers to swallow. Daniel Dennett, a champion of evolution whose applications of Darwinian logic to biological questions I much admire, immediately recognizes the danger.

> Believers in any of the proposed strong versions of the Anthropic Principle think they can deduce something wonderful and surprising from the fact that we conscious observers are here—for instance, that in some sense the universe exists *for* us, or that perhaps we exist *so that* the universe as a whole can exist, or even that God created the universe the way He did so that we would be possible.[12]

But don't think for a moment that Dennett, a vocal opponent of religion, is going to cede any ground on this point. He does have an alternative in mind, but the length to which he must go in order to prop up that alternative shows exactly how much he fears the anthropic principle.

We live, so far as we can tell, in just one universe, but what if there were more? What if the forces of nature were somehow capable of spinning out multiple, independent universes? Better still, what if the physical constants of each of those universes were slightly different? We'd then have, in effect, a way to generate, given unlimited time and space, every conceivable kind of universe, including our own.

Dennett eagerly embraces the idea of multiplying universes, because he sees them as a way to contain the looming theological problems posed by the anthropic principle. He notes that physicist Lee Smolin has suggested a cosmology in which black holes are the birthplaces of alternate universes. This means that the black holes in our own universe might be, in a cosmological kind of way, the organs of reproduction from which new universes emerge. The fundamental constants of each of a black hole's "offspring" universes are slight variations of those found in the "parent" universe, and for Dennett, this is a way out of the anthropic trap.

Using Smolin's ideas as the engine to produce a nearly infinite number of universes, Dennett reasons that sooner or later—and quite accidentally—in one of those universes the combination of laws and constants would be just right for the evolution of life, and bingo: a couple of billion years later, beings like us appear. Eager, curious, and annoyingly scientific. We wouldn't know anything about all those other universes, and we'd probably conclude that our universe had been tailored just for us.

In Dennett's view, these multiple universes have undergone a kind of Darwinian natural selection—only a few of them had physical constants that were ultimately capable of supporting life, and those are the only ones in which living observers could have evolved. You and I, living in just such a universe, find our good fortune amazing, but Dennett knows it's nothing of the sort. Though the person with a winning lottery ticket believes himself to be incredibly lucky, the organizer of the lottery knows very well that one person will always win the prize. To the lottery broker, the emergence of a winner is a certainty, not a stroke of luck.

The same logic, incidentally, explains why evolution is able to "find" a working solution for so many problems of physiology and biochemistry. Looking backwards, the path that evolution has taken to produce *any* individual result, any protein, organ, or species, looks remarkably lucky, but the reason for such great "luck" is the fact that evolution throws so many different variations into the wind for natural selection to sort out. The ability of organisms to generate, shuffle, and test multiple versions of their genetic information is the reason why evolution does not require a programmer to get things right the first time.

Dennett figures that he can apply the same approach to account for the pleasant convergence of physical constants that makes life possible. Evolution works on a multitude of genetic variations in a population of organisms, so he reasons that there could be a multitude of universes, each with varying physical constants. The problem, of course, is that we *know* that organisms reproduce. But universes? Dennett knows that we will never be able to find, even in principle, evidence for any of those parallel universes. If they existed, we could neither communicate with them nor observe them. Nonetheless, he is willing to postulate their existence because it relieves us of the need to find another reason for the elegant "anthropic coincidences" of our universe. To those who doubt his solution, he writes that a multiplying swarm of universes is at least as good an explanation "as any traditional alternative."[13]

Dennett's willingness to advance Smolin's replicating universe idea in a book ostensibly on Darwinian evolution shows the importance he attaches to finding an answer to the anthropic argument. It also, however, has the unintended effect of revealing the strength of that argument. Remember that Dennett holds this idea up as his only counter-proposal to the "traditional alternative." Human beings, even those educated in the scientific tradition, have begun to wonder why the universe is organized in a way that is distinctly hospitable to life. This is an important question, and Dennett realizes that it demands an answer. He's found one, he tells us, in the multiple universe hypothesis. Unfortunately, he must also admit that we may never, even in principle, find evidence for another universe. Deprived of empirical evidence, the best he can manage is to assure his readers that his non-theistic explanation is "at least as good" as a theistic one. The other side of that coin may not be apparent to Dennett. By writing that his explanation is just as good as its "traditional alternative," Dennett unwittingly admits that the "traditional alternative" is valid as well.

The traditional alternative, of course, is God. Even as we use experimental science and mathematical logic to reveal the laws and structure of the physical universe, a series of important questions will always remain, including the sources of those laws and the reason for there being a universe in the first place.[14] Before we began this journey, we knew that we existed, and that alone should have told us, from the beginning, that our universe was one that made life possible. Therefore, we should not be even slightly surprised to learn that the physical constants of gravity, electromagnetism, and the nuclear forces are compatible with life. Nonetheless, if we once thought we had been dealt nothing more than a typical cosmic hand, a selection of cards with arbitrary values, determined at random in the dust and chaos of the big bang, then we have some serious explaining to do.

Just Sheer, Dumb Luck?

A little side trip into the constants of the universe may be invigorating, but let's keep in mind that evolution is a *biological* theory, not a cosmic one. The success or failure of the anthropic principle may be relevant to whether or not we can find God in the stars, but it does not tell us whether or not we can find Him in the evolution of species. To do that, we have to look directly at the events that drive evolution.

The explanatory power of evolution derives from its simplicity. Natural selection *favors* and *preserves* those variations that work best, and new variation is constantly generated by mutations, gene rearrangements, and even by exchanges of genetic information between organisms. This does not mean that the path of evolution is random in the sense that *anything* can happen as we jump from one generation to the next. Although the potential for change is always constrained by realities of physiology, the demands of the environment, and the current complexities of the genetic system, the input of variation into any genetic system is unpredictable. Although not completely random, chance does affect which mutations, which mistakes, appear in which individuals. As we saw earlier, this inherent unpredictability is not a matter of inadequate scientific knowledge. Rather, it is a reflection that the behavior of matter itself is indeterminate, and therefore unpredictable. It is one of the reasons why we cannot predict, with any detailed certainty, the future path of evolution.

For many religious people, here lies the problem with evolution. No matter how much experimental or historical evidence can be marshaled to support it, evolution is still a chance, random process. Doesn't the very randomness of evolution rule out any notion of divine purpose? More to the point, if mankind is the *intentional* creation of God, as all Western religions teach, how could He possibly have used evolution to fashion the very creatures He made in His own image and likeness?

Did God rig the evolutionary process to give rise to us? Did He only make things *look* natural, allowing us to emerge at the long end of a chain of events that might look random, but really aren't? And if He did, why? Why would He have gone to such trouble to make our appearance seem to be the result of a random, natural process when He could have simply brought us into existence in a flash? This is an aspect of evolution that has bothered believers for a long time. Even Darwin struggled with the issue, exchanging letters on the subject with Harvard botanist Asa Gray, one of his strongest scientific supporters but also a devout Christian. In March of 1860, Gray wrote a strongly positive review of *The Origin* in the *American Journal of Science,* but privately he wrote to Darwin expressing concern about its implications for Christianity. Gray supported evolution and its mechanism, natural selection, but he wondered how a natural world ruled in this way could leave room for the divine. Darwin's glum reply was of little help, allowing only that he was "inclined to look at everything as resulting from designed laws, with the details, whether good or bad, left to the working out of what we may call chance."[15]

There, in a nutshell, is the problem. If evolution really did take place, then God must have rigged everything. Otherwise, how could He have been sure that evolution would have produced *us*? And if He didn't, isn't our species, *Homo sapiens,* nothing more than, as Gould put it, a "tiny twig on an improbable branch of a contingent limb on a fortunate tree?"[16] Surely we could not be *both* the products of evolution and the apple of God's eye?

Well, in fact we could. The great problem posed by evolution's indeterminacy is more apparent than real.

To appreciate this, let's step back from the specific issue of evolution and ask a more basic question: Is chance real? Are there events in nature whose outcomes are genuinely uncertain? In physics the answer is a resounding yes, but we don't have to aim photons through diffraction slits to confirm the reality of chance. A state lottery, properly conducted, is a genuine chance event. We speak of "chance encounters," the "luck of the draw," being in the "wrong place at the wrong time." A tornado destroys the houses on one side of a street, but leaves the other side untouched. One soldier survives a bombardment, but another is killed by shrapnel. The fragments of an airplane, shattered by a bomb, crash down upon a handful of people in an unsuspecting community. Most on the ground are spared, but a few are not. Are all these events matters of chance? In ordinary terms, they certainly are. There's a reason for the wry saying that it's better to be lucky than good.

For a person who believes in God, how do such chance events fit into the scheme of things? One might say that chance is an illusion—that actually, everything is controlled by the hand of God. You're going to win the Megabucks tonight if God wants you to, and if He doesn't, you might just as well not buy a ticket. When a stray bullet is randomly fired into a crowd, God decides who will get hit, and who will survive. When the biggest hurricane of the year hits Key West and not Miami, it had to be God's will. You could, I suppose, cast the Almighty in this guise, make Him a cosmic tyrant, a grand puppeteer pulling every string at once, and then nothing would be left to chance. But most people would find this view of God disturbing. Putting God in charge of every trip and stumble of our daily lives does take chance out of the picture, but at what price? God is now personally responsible for falling limbs and power lines, for your daughter's illness, and even for the school bus full of children slipping off an icy road.

A different view, one widely held by Western religions, is that chance events are genuine because the physical world has an existence independent of God's will. When you and I flip a coin to decide who gets to eat that last slice of pizza, the outcome is not known in advance—God really doesn't care who has already eaten his share of pepperoni. Chance is not only consistent with the idea of God, it is the only way in which a truly independent physical reality can exist.

Even deeply religious people can fully appreciate the role that chance can play in human history. During the Second World War, the Japanese destroyer *Amagiri* literally sliced a tiny American boat in half, killing two of the thirteen crewmen on board. The collision took place at 2 A.M. in near-total darkness. Had her commander, Lieutenant John F. Kennedy, been near the center of the PT-109 at the moment of impact or had the bow wave from the *Amagiri* not doused the flames in the section of the boat into which Kennedy was flung, the political history of the United States would have been quite different. The contingent nature of history routinely enlarges simple matters of chance into events of historical significance. Retreat far enough into the past and you will find that individual choices and chance events of seemingly minuscule importance have reversed the tides of history.

The same is certainly true of biological history, a theme explored elegantly in Gould's book *Wonderful Life*. Considering the events in natural history that led to our own emergence on this planet, we can ask whether events leading to the evolution of the human species *had* to come out the same way? Did the ancestors of vertebrates *have* to survive the Cambrian? Did mammals *have* to evolve from the vertebrates? Did one group of mammals, the primates, *have* to take to the trees? Was one tiny African branch of these tree climbers absolutely predestined to survive and give rise to *Homo sapiens*? The answer in every case is no.[17] Those events didn't *have to* come out that way, any more than Lieutenant Kennedy was certain to survive a collision with a ship fifty times larger than his. As time goes by, in any historical process, human or biological, contingency casts smaller and smaller events, many governed by chance, into fate's most important roles.

Does this leave no room for God? Does this mean, as Gould asserts, that we cannot view our species as the handiwork of the Creator? Not at all.

Let's make it personal. Just about every preacher I have had the patience to listen to has eventually said that God has a plan for each of us. That God wanted you and me to exist. That God knew us and loved us even before we were born. That idea is consonant with all of the great Western religions. Should I be required to reject it if can be shown that chance played a role in making me who I am?

Chance did, of course. As a biologist, I am perfectly well aware that important aspects of the sorting of chromosomes, a process called "meiosis" that takes place when reproductive cells are formed, are random—matters of genuine, blind chance. That means that the person who I am, the selection of genes from my mother and father that determined my physical characteristics, was very much a matter of chance. For that matter, so was the fact that my parents ever met, brought together by the great calamity of World War II, a soldier boy from Indiana smiling at a pretty girl from New Jersey at a USO dance near Fort Dix.

Since chance—and for that matter, free will—played such an important role in bringing me to life, does this mean that I cannot view my own existence as part of God's plan? Of course not. Any clergyman, very much in the Christian tradition, would caution me that God's purpose does not always submit to human analysis. God's means are beyond our ability to fathom, and just because events seem to have ordinary causes, or seem to be the result of chance, does not mean that they are not part of that divine plan. This is the reason why no religious person would take issue with a geneticist's assertion that the sorting of chromosomes in meiosis is random. Sure it is, every bit as random as the flipping of a coin, the impact of a meteor, or a sudden shift in climate that drives one species to extinction and allows another to survive.

A Christian, specifically, sees his life, his family, and his small place in history as parts of God's plan. He has faith that God expects him to use his talents and abilities in God's name. He accepts the adversity that comes into his life as a challenge from God, and he sees apparent misfortune as an opportunity to do good in the service of both God and man. These noncontroversial elements of Christian teaching are so ordinary that we sometimes forget what they imply about the interplay of history, free will, and chance. To put it simply, they mean that God, if He exists, surpasses our ordinary understanding of chance and causality. Christians know that chance plays an undeniable role in history, and nonetheless accept the events that affect them in their daily lives as part of God's plan for each of them. This means that Christians *already* agree that the details of a historical process can be driven by chance, that to allow for individual free will the outcome of such a process need not be preordained, and that the final result of the process may nonetheless be seen as part of God's will. These ordinary elements of religious teaching merge smoothly into everything we know about evolution.

Biological history, just like human history, is a contingent process. The chance extinction of a rare species can profoundly affect the course of evolution in the same way that the unexpected death of a single person could change political or economic history. Wind the tape of history back to the era of the Civil War. Maybe this time Pickett's charge succeeds. Maybe this time the bullet from Booth's gun fails to kill the President. Maybe this time an ancestor of Kennedy or Nixon or Reagan does not recover from his wounds, and a political figure does not appear four generations hence. Allow any of these chance events another roll of the dice, and the twentieth century could easily have been very different—the next century more different still.

On earth in the year 2000, the most important single nation, in economic, political, and military terms, is the United States of America. Would it be unfair to say that no one could have predicted this a thousand years earlier? Of course not. Was the establishment of the American nation and its rise to power inevitable? Hardly. The historical position of any country comes about as the result of choices made by millions of people, by the actions of those who rise to power in many nations, and by their interplay with the outcomes of a host of fundamentally indeterminant events. History itself is an unpredictable process, and it need not have turned out the way it did. That much is almost self-evident. The question for us now is whether the inherent unpredictability of history should lead any person to conclude that God could not have used that historical process to produce the world in which we live today. I submit that the answer is no, that the ebb and flow of human history is entirely consistent with the Western conception of God.

Evolution answers the question of chance and purpose in *exactly* the same way that history answers questions about the course of human events. To a biologist, evolution is subject to chance and unpredictability, just like human history. Its outcome is uncertain, and likely to be unrepeatable, just like human history. And evolution admits to no obvious purpose or single goal, just like human history. History, like evolution, seems to occur without divine guidance. No one seems to think that a religious person engaged in the study of history must find a way that God rigged human events in order to cause the Civil War, the Industrial Revolution, or the Holocaust. Yet curiously, that is exactly what many expect of a religious person engaged in the study of natural history—they want to know how God could have ensured the success of mammals, the rise of flowering plants, and most especially, the ascent of man.

My answer, in every case, is that God need not have. Evolution is not rigged, and religious belief does not require one to postulate a God who fixes the game, bribes the referees, or tricks natural selection. The reality of natural history, like the reality of human history, is more interesting and more exciting.

The freedom to act and choose enjoyed by each individual in the Western religious tradition requires that God allow the future of His creation to be left open. This leaves the future accessible to human choice, and to the multitude of possibilities that we ourselves create. If events in the material world were strictly determined, then evolution would indeed move towards the predictable outcomes that so many people seem to want; but if this were the case, how could the future truly be open? As material beings, our actions and even our thoughts would be preordained, and our freedom to act and choose would disappear within a tangle of biochemical machinery. But events in the material world are not strictly determined, the outcome of evolution is not predictable, and as a result, our freedom to act as independent beings is preserved.

So, how is a random, chance process like evolution consistent with the will of a Creator? As physicist and religious scholar Ian Barbour, the author of a dozen books on science and religion,[18] puts it:

> Natural laws and chance may equally be instruments of God's intentions. There can be purpose without an exact predetermined plan.[19]

How is this possible? I would submit that if we can see the hand of God in the unpredictable events of history, if we can see meaning and purpose in the challenges and trials of our daily lives, then we can certainly see God's will emerging in the grand and improbable tree of life.

Would God's purpose have been realized if evolution had turned out a little differently? How can we say for sure? But this much I think is clear: Given evolution's ability to adapt, to innovate, to test, and to experiment, sooner or later it would have given the Creator exactly what He was looking for—a creature who, like us, could know Him and love Him, could perceive the heavens and dream of the stars, a creature who would eventually discover the extraordinary process of evolution that filled His earth with so much life.

Finding Work

Obviously, few religious people find it problematical that their own *personal* existence might not have been preordained by God, that they might not be here but for the decisions of their parents or the chance events that brought them together. But strangely, some of the very same people find it inconceivable that the *biological* existence of our species could have been subject to exactly the same forces. If we can see God's will in the flow of history and the circumstances of our daily lives, we can certainly see it in the currents of natural history.

A similar argument applies to another concern—the notion that evolution, making nature self-sufficient, leaves nothing for God to do. It becomes necessary—a theological imperative—to find some element of the natural world that *cannot* be explained by scientific materialism. If we find such elements, we can rescue God from the ranks of deist unemployment, making Him active and necessary once again—or so the thinking goes.

As we saw in Chapter 7, the adherents of this viewpoint always feel constrained to find events in the *past* for which the direct action of a designer is invoked, even though they steadfastly maintain that God acts in the present. That's exactly how He acted in the past—should not God's will and His means of executing it be consistent through the ages?

The Christian God isn't a deist one; neither is Allah, or the God of Abraham. Any God worthy of the name has to be capable of miracles, and each of the great Western religions attributes a number of very specific miracles to their conception of God. What can science say about a miracle? Nothing. By definition, the miraculous is beyond explanation, beyond our understanding, beyond science. This does not mean that miracles do not occur. A key doctrine in my own faith is that Jesus was born of a virgin, even though it makes no scientific sense—there is the matter of Jesus's Y-chromosome to account for. But that is the point. Miracles, by definition, do not have to make scientific sense. They are specific acts of God, designed in most cases to get a message across. Their very rarity is what makes them remarkable.

An atheist, of course, will argue that miracles are bogus and are stitched together from the needs and desires of human imagination. Maybe so. To a believer, miracles are more than "violations of the laws of nature," as they were once described by David Hume. They reflect a greater reality, a spiritual reality, and they occur in a context that makes religious, not scientific, sense. To the nonbeliever, there is no spiritual reality, and hence there are no miracles. To a person of faith, miracles display the greater purposes of God, giving them a meaning that transcends physical reality.

If this is true, why shouldn't we allow that the creation of our species was a miracle? Or why not agree that the sudden explosion of life in the Cambrian era might have been a miracle? Both might have been. In 1900, we could easily have said that the sun's fire was a miracle. Unable to explain the biological basis of immunity, we could have chalked that up to God, too. And for good measure we could have told our students that the interior heat of the earth might be the work of the devil.

We are now far enough along in the development of science to appreciate that its track record suggests that ultimately it will find natural causes for natural phenomena. God has fashioned a self-consistent reality in nature, and He allows us to work within it. This may be frustrating to those who look for signs and wonders, but it's clearly the way life is. Comedian Bill Cosby's wonderful "Noah" routine emphasizes this point brilliantly. God's voice comes from above, instructing Noah to build an ark. Cosby's Noah, more south-side Philly than biblical, snickers, "Who is this, *really*?" The voice echoes back, "This is God!" And Noah answers, "Yeah, right. Am I on Candid Camera?" A bit annoyed, God establishes His identity and gets to the business of instructing Noah on the details of the ark. But Noah, a little less than eager to construct this huge boat, asks the question that must have occurred to every child who has ever read the biblical passage: "If you want this ark, why don't you build it yourself?" "Noah," God's voice thunders, "you know I don't work that way!"

Exactly. God's miracles are not routine subversions of the laws of nature. If they were, then the issue of why so many extinct forms of life preceeded us would be a conundrum, since each one would have to be the intentional creation of the mind of God. If each were just another chapter in an unfolding plan driven by the laws and principles of nature, the issue is not important. If God were just a magician, He could have made the present world appear in a puff of smoke. But He isn't. As the good Dr. Cosby has pointed out, we *already* know He doesn't work that way!

Theologically, the care that God takes *not* to intervene pointlessly in the world is an essential part of His plan for us. Ian Barbour claims that human freedom would be impossible without God's willingness to limit His actions.

> If all power is on God's side, what powers are assignable to humanity? . . . But if omnipotence is defended, and everything that happens is God's will, then God is responsible for evil and suffering and God's goodness is compromised.[20]

This does not mean, as Barbour is careful to explain, that God cannot or does not act in the world. It simply means that He is wise enough to act in ways that preserve our own freedom, allowing us to reap the rewards and consequences of our own free will.

Even the most devout believer would have to say that when God does act in the world, He does so with care and with subtlety. At a minimum, the continuing existence of the universe itself can be attributed to God. The existence of the universe is not self-explanatory, and to a believer the existence of every particle, wave, and field is a product of the continuing will of God. That's a start which would keep most of us busy, but the Western understanding of God requires more than universal maintenance. Fortunately, in scientific terms, if there is a God, He has left Himself plenty of material to work with. To pick just one example, the indeterminate nature of quantum events would allow a clever and subtle God to influence events in ways that are profound, but scientifically undetectable to us. Those events could include the appearance of mutations, the activation of individual neurons in the brain, and even the survival of individual cells and organisms affected by the chance processes of radioactive decay. Chaos theory emphasizes the fact that enormous changes in physical systems can be brought about by unimaginably small changes in initial conditions; and this, too, could serve as an undetectable amplifier of divine action.

Another opportunity for divine action can be found in time itself. John

Polkinghorne, a distinguished theoretical physicist who is also an Anglican priest, has pointed out that if Western theologians agree on anything, it is on the eternal nature of God.[21] This means that God, who always has been and always will be, transcends time and therefore is the master of it. We, on the other hand, traverse time in a linear fashion. An eternal being who is present everywhere and at all times could easily act to alter what both physicists and Hollywood call the space-time continuum in ways that profoundly affect events. Locked into a single point in time and moving in a single, unchanging path from the present to the future, we wouldn't have a clue. And God, the Creator of space, time, chance, and indeterminacy, would exercise exactly the degree of control He chooses.

Finally, any traditional believer must agree that God is able to influence the thoughts and actions of individual human beings. We pray for strength, we pray for patience, and we pray for understanding. Prayer is an element of faith, and bound within it is the conviction that God can affect us and those for whom we pray in positive ways. With great effort, I suppose we could break the possibilities for God's actions within us into cellular and biochemical terms. We would then find that the same arguments regarding indeterminacy and chaotic systems can be applied to individual neurons, the activities of which are the physical basis of our thoughts and feelings. God could, in principle, act without being scientifically detectable, but even within us, He would still face the same problems of agency and free will that confront any divine action. No religious person believes that God ever acts so decisively as to force us to do good, or compel us to obedience. The common experience of religious people is that God provides assistance, inspiration, and strength—but to accept those gifts still requires an act of human will, a free choice to do what is right despite the burden and even the suffering that may result.

John Polkinghorne puts it this way:

> The actual balance between chance and necessity, contingency and potentiality which we perceive seems to me to be consistent with the will of a patient and understanding Creator, content to achieve his purpose through the unfolding of process and accepting thereby a measure of the vulnerability and precariousness which always characterize the gift of freedom by love.[22]

By any reasonable analysis, evolution does nothing to distance or to weaken the power of God. We already know that we live in a world of natural causes, explicable by the workings of natural law. All that evolution does is to extend the workings of these natural laws to the novelty of life and to its changes over time. A God who presides over an evolutionary process is not an impotent, passive observer. Rather, He is one whose genius fashioned a fruitful world in which the process of continuing creation is woven into the fabric of matter itself. He retains the freedom to act, to reveal Himself to His creatures, to inspire, and to teach. He is the master of chance and time, whose actions, both powerful and subtle, respect the independence of His creation and give human beings the genuine freedom to accept or to reject His love.

Taking the Scenic Route

"I know that God didn't use evolution to produce us," the parent of a high school student once told me, "because it would have taken too long." If God's ultimate purpose was to produce human beings, then why would He have fooled around with a process that took billions of years? As easy as it might be to answer an argument based simply on the notion that God (who is eternal) must have been in a hurry, let's extend it a bit. Gould has fashioned similar arguments based on his analysis of the nature of evolutionary change over the epochs of geological time. He believes that a proper understanding of natural history undermines any species-based conceit that animals stand at the inevitable pinnacle of life:

> But the real enigma—at least with respect to our parochial concerns about the progressive inevitability of our own lineage—surrounds the origin and early history of animals. If life had always been hankering to reach a pinnacle of expression as the animal kingdom, then organic history seemed to be in no hurry to initiate this ultimate phase. About five sixths of life's history had passed before animals made their first appearance in the fossil record some 600 million years ago.[23]

We're animals. And if animal life was to be the chosen vessel from which its ultimate expression, humanity, was to spring forth, then why did most of the earth's history pass without it? If the 4.5 billion years of earth's history were represented as a single twenty-four-hour day, then animals would appear only in the last three hours and fifteen minutes. According to Gould, that seems pretty late if animals were ultimate and inevitable. If the purpose of the whole process was to produce our species, as my friend implied, shouldn't we be bothered by the fact that human beings arose only in the last thirty-eight seconds?[24]

If you find this reasoning appealing, keep going. Even the immensities of geological time dwindle against the age of the universe. How are the heavy elements of the universe produced? By 10 or 15 billion years of cooking in the interiors of stars, which must then collapse and explode to throw them into space where they become available to form secondary stars, like our sun; and planets, like the earth. Now we really look like an afterthought, don't we, if indeed we are a thought at all!

In the view of my friend, who *knows* that creating us was God's intention from the beginning, such facts are enough to rule out evolution itself. But why should they? Evolution is a natural process, and natural processes are undirected. Even if God can intervene in nature, why should He when nature can do a perfectly fine job of achieving His aims all by itself? It was God, after all, who chose the universal constants that made life possible. The notion that God had to act quickly and directly to produce us contradicts not only the scientific evidence of how our species arose, but even a strictly theological reading of history. Why should so many nonhuman species have preceded us on this planet? Good question. We might just as well ask why so many human civilizations preceded our own. We could ask why God, if He was interested in all the peoples of the world, first revealed Himself only to a few desert tribes in the Middle East. And if He was interested in redeeming *all* people from their sins, why did He allow scores of generations to pass unsaved before He sent His divine son?

If we are sure enough of the Creator's thoughts to know that He would have made our species by a more direct route, then what are we to make of the highly indirect routes that led to our modern civilizations, our languages, and even our personal lives? What was God's purpose in allowing evolution to produce the great dinosaurs of the Jurassic? Beats me. But I am equally clueless with respect to God's purpose in the Mayas, the Toltecs, and the other great civilizations of the Americas that rose and fell without ever being exposed to His word. To demand that a person of faith account for each and every event in human history or natural history is to demand the impossible—that they know every detail of God's actions.

Whatever God's characteristics, impatience is not one of them. A genuine believer trusts not only in the reality of the divine, but also in its wisdom, and sees history, indirect, comic, and tragic, as the unfolding of His plan. There is no religious justification for demanding that the Creator hold to a certain schedule, follow a defined pathway, or do things exactly according to our expectations. To God, a thousand years are as a twinkling, and there is no reason to believe that our appearance on this planet was for Him anything other than right on time.

No More Mr. Nice Guy

Of all the concerns expressed by Christians with respect to evolution, the strangest, the least logical, the most bizarre is the idea that evolution is too cruel to be compatible with their notion of a loving God. In this view, the Darwinian concept of natural selection, a struggle for survival, puts the weak at the mercy of the strong. It rewards the ruthless and punishes the kind. It leads to "nature red in tooth and claw," and could never have produced the world of harmony, beauty, and balance that surrounds us.

Darwin himself was worried about this, and tried to assuage Victorian sensibilities with a few kind words placed at the conclusion of the third chapter of *The Origin*:

> All that we can do, is to keep steadily in mind that each organic being is striving to increase at a geometrical ratio; that each at some period of its life, during some season of the year, during each generation or at intervals, has to struggle for life, and to suffer great destruction. When we reflect on this struggle, we may console ourselves with the full belief, that the war of nature is not incessant, that no fear is felt, that death is generally prompt, and that the vigorous, the healthy, and the happy survive and multiply.[25]

Could evolution really be so cruel as to require such an apology? To answer that question we have to keep two things in mind. The first is that cruelty is relative. As a New Englander, I enjoy few things more than a lobster dinner, especially at the end of a long summer day. The preparation of that meal, from the point of view of the lobster, is an act of unmitigated cruelty, perpetrated by me, the cook. The cats patrolling our barn are lovable family pets, but I assure you that their actions in keeping that barn vermin-free meet the highest standards of viciousness. Like beauty, the brutality of life is in the eye of the beholder.

The second is that we cannot call evolution cruel if all we are really doing is assigning to evolution the raw savagery of nature itself. The reality of life is that the world often lacks mercy, pity, and even common decency. If you doubt that, read Jack London's short story "To Build a Fire," and think of yourself freezing to death; or better still, spend some time watching the operations at a well-run commercial slaughterhouse. You might even try what Darwin did one spring, to systematically record each individual plant in a patch of garden as it sprang to life. You could then follow them throughout the growing season. After a few months, you would know how many of them survive to reproduce. The answer: very few. Even an English garden is subject to the law of the jungle, a law that affects our species, too. Ironically, anyone who believes that evolution introduced the idea of systematic cruelty to human nature is likely overlooking many acts of biblical cruelty, especially God's killing of the Egyptian firstborns, Herod's murder of the innocents, or the complete and intentional destruction of the cities of Sodom and Gomorrah.

Evolution cannot be a cruel concept if all it does is reflect the realities of nature, including birth, struggle, life, and death. It is a fact—not a feature of evolutionary theory, but an objective reality—that every organism alive will eventually die. Some will leave offspring and some will not. And from those two facts come the forces that drive and control natural selection. Darwin immediately focused on the obvious—that *competition* exists among individuals for food, for resources, and for mates. As a result, evolution is generally cast as a struggle for existence in which the interests of the individual are paramount. Greed, self-interest, and exploitation may be cardinal virtues in such a world, but here is where the value of Darwin's work is especially clear.

His simple and durable framework held the key for explaining something that seems most undarwinian, something that baffled Darwin himself—altruism. In a world that ought to be filled with vicious competition, why are there so many examples of animals being *nice* to each other? Why does a bird protect her nest, her offspring with her very life? Why do caribou form herds in which the strongest guard the youngest from advancing predators? Why do worker bees enthusiastically end their own lives to sting, kamikazi-style, any threat to their hives? Why do worker ants, those heroes of socialist labor, eagerly share the food they have gathered with all the members of their colony? Shouldn't self-interest dictate, in every case, that these altruists, these brave individuals, would be better off ignoring their fellows and looking out for themselves? Doesn't evolution say that the only thing that matters is looking out for number one? As it turns out, it doesn't.

One of the most recent additions to evolutionary theory has come from the field of sociobiology, which studies the biological basis of social behavior. Although mammals and birds can learn behaviors, in most other animals, especially the insects, behaviors are instinctive—they are inherited. A female mosquito finds your exposed skin even in total darkness, because her genes have programmed her to follow—instinctively—chemical signals in the vapors of carbon dioxide and moisture that rise from your body. In her case, the genes for that behavior are favored by natural selection because they increase her chances of getting a meal, enabling her to lay more eggs, and increasing the chances that she will have lots of offspring in the next generation. The fact that such behaviors are inherited means that instinctive behavior can be shaped by natural selection, just as surely as any physical trait.

Darwin appreciated that fact, and he figured that natural selection would favor behaviors that helped to find food, attract mates, and provide for self-defense, all of which it does. But what about altruistic behavior? What about self-sacrificing behavior? What about cooperative behavior? Shouldn't genes for these *nice* behaviors, which help *other* individuals, lose big in the competition against nasty, ruthless, *self*-promoting behaviors? As farsighted biologists like Edward O. Wilson have shown, the answer is, not necessarily.

Altruistic behaviors in nature are most often directed towards helping close relatives—an animal's offspring, its cousins, sometimes its siblings. The unique genetic systems of social insects guarantee that nearly all the members of the colony are even more closely related than full brothers and sisters.[26] Why does this matter? For the very simple reason that if a gene produces altruistic behaviors that are directed towards close relatives, there is a good chance (fifty percent in the case of parent and offspring) that the gene is actually helping a copy of itself to survive. The implications of this simple statement, which were wonderfully explored in Richard Dawkins's book *The Selfish Gene,* are profound. Darwinian evolution can produce cooperation and care just as surely as conflict and competition. The care and self-sacrifice seen in animal families are not exceptions to evolution—rather, they are the straightforward results of natural selection acting to favor instinctive altruism. Under the right circumstances, nice guys really do finish first. Once again, Darwin got it right.

Evolutionary success, it turns out, is rooted a little deeper than shallow self-preservation. To be sure, natural selection produces speed and cunning, claws and teeth, but it also fills the world with a beauty that is almost beyond description. Part of that beauty can be found in the loving care of a mother for her young, in the exuberant play of packs and herds, and even in the signs and signals by which the tiniest of insects build their nests and work the soils of the earth.

As a biologist, I confess to viewing nature with a certain passion. What I see in life, and what is seen by others who share that passion, is for sure a little rough around the edges. Life is never tidy, never restrained, never willing to behave according to our rules. In the eyes of those who recoil from the realities of nature, this no doubt means that the hard lessons she would teach must be resisted or denied. There are greater lessons, in my view, greater poetry to be learned from a world of harsh songs and terrible beauty. Annie Dillard, author of *Pilgrim at Tinker Creek,* put it perfectly:

> The wonder is—given the errant nature of freedom and the burgeoning of texture in time—the wonder is that all the forms are not monsters, that there is beauty at all, grace gratuitous, pennies found, like mockingbird's free fall. Beauty itself is the fruit of the creator's exuberance that grew such a tangle, and the grotesques and horrors bloom from the same free growth, that intricate scramble and twine up and down the conditions of time.[27]

Seeing It His Way

As a scientist, I find it easy to answer most questions about evolution. We know the age of the earth, we have found an abundance of the record of past life to document descent with modification, and we can observe rapid evolutionary change in the lab and in the field with our own eyes. Despite all of this, there remains one objection I can never answer as a scientist, and I've heard it many times: "*My* God wouldn't have done things that way." The personality of God, in the eyes of such a skeptic, is such that He would find evolution distasteful.

There is no scientific answer to this objection—how can science possibly respond to an argument built around the *personality* of the Creator? How can we even conceive of scientific evidence that might tell us anything about the attitude of God, and whether or not it is compatible with evolution? I have already argued that science cannot give us a definitive answer as to whether or not God exists; and if that question is out of bounds for science, how can we possibly expect it to tell us what sort of fellow He may be?

Obviously, science isn't going to satisfy anyone who makes the "argument from personality."

Starting not as a scientist, but simply as a believer, I will address the question this way: Is what science tells us of the physical world, including evolution, compatible with what we *think* we know about God? In other words, if we try to put ourselves in God's shoes, could evolution have been part of His plan of creation?

Ultimately, it is impossible to answer this question. We cannot know everything God had in mind, but we can know this about our *beliefs of God:* All of the Western monotheistic religions maintain that God brought the universe into being, that He intended to create creatures deserving of a soul, and that He wished that universe to be a place in which those creatures had a truly free choice between good and evil, between God and darkness. Given those theological basics, let's see where we can go.

Our first step would be to assume that an all-powerful Deity decided to make creatures and to endow them with free will and the ability to make moral choices. He loved these creatures, but He did not wish to place them into the same spiritual reality that He Himself occupies. Instead, He constructed a material universe in which these creatures would exist, upon which their lives would depend. As science tells us, that dependence is real, not an illusion. The Creator could, I suppose, have commanded stones to live. He could have made creatures whose substance was no different from the inanimate objects around them. Being intelligent (as was His wish), they eventually would have discovered that their composition was no different from that of nonliving objects. The direct dependence of their animus, their life, upon the will of God would then have been obvious. But if that was His goal, then why make a distinct material world at all? The genius of the Creator's plan was that by creating a separate world, a world that ran by its own rules, He would give His creatures the "space" they would need to become independent, to make true moral choices.

Although we might like to think there is a divine spark within us, in fact, our lives and the lives of every animal, plant, and microorganism of this planet have a physical basis in the chemistry of cells and tissues. We share a physical continuity with the lives of others, and are inextricably linked to all life on earth. New individuals do not spring, like Athena, from the minds of gods, but rather are born of the flesh, from the union of sperm and egg. No question about it—our origins as individuals come entirely from the materials of life. The Creator fashioned a world in which matter became the basis of life.

In constructing this physical world, the Creator faced a practical problem, not unlike the problem facing a zoologist who wishes to provide a wild animal with something close to its natural habitat. How should the matter of this physical world behave, what rules should it follow? If the Creator made the behavior of this material world logical and sensible, He would ultimately allow His intelligent creatures to investigate the universe and discover the properties of the matter He has created. If the behavior of that matter, at all levels, were to be governed by laws making the outcomes of all natural processes inherently predictable, then the entire structure of that universe would be a self-contained and self-sufficient clockwork. Two profound problems would result. The Creator would be unable to intervene, except by suspending the laws of His world in a visible way. And what is perhaps more important, His creatures could not have the freedom He desired for them. How could they, if they were only machines, made up of bits and pieces of matter following precise laws? Worse still, they might eventually figure that out.

The only way He could avoid His creatures' eventual discovery of their own determinant clockwork would be to limit their ability to question, learn, and discover—something that He clearly would not want to do if He wanted these creatures to be free to choose Him or to reject Him. So, what could the Creator do? Exactly these problems led many of the best minds of the eighteenth and nineteenth centuries to reject God as a logical impossibility, but it turned out that our Creator had a brilliant solution to all of these problems.

His solution was to fashion a material universe in which the conditions of precise determinism do not apply. On the larger scale, He made the averaged behavior of matter sensible and predictable, making it possible to construct organisms and environments that work according to natural laws, which His creatures were sure to discover. Then He ensured that the behavior of the material fine structure of the universe was inherently unpredictable due to the multiple effects of quantum indeterminacy. This would allow His creatures to develop a science that applies to large-scale interactions, but one that is forever forbidden from grasping the detailed behavior of the units from which the material world is fashioned. That, of course, is exactly the world in which we live.

To be sure, there is nothing about quantum indeterminacy at the base of our existence that proves it was the work of God. Or of a design to the universe clearly put there by an intelligent force to accommodate living things. However, if there is a God, consider what a master stroke quantum indeterminacy was. To create an orderly material world that didn't require constant intervention, the Creator *had* to make things obey defined laws. But if those laws were to run all the way down to the building blocks of matter, they would also have denied free will. They would have made it possible for His creatures (eventually) to figure out that all past events and all future ones could be inferred from a single reading of the state of the physical world at any given time.

Remarkably, what quantum indeterminacy does is to *deny* us the possibility of that ever happening. We cannot uncritically extrapolate the details of the present backwards to learn the past; and the future is what we make of it. Were this not the case, the future would be what our particles make of us. Instead, we are inextricably locked into the present, with our thoughts, words, and deeds helping to construct the future, a future that remains open to our own choices, to a world of possibilities.

Once He had fixed the physical nature of our universe, once He had ensured that the constants of nature would create a chemistry and physics that allowed for life, God would then have gone about the process of producing the creatures that would share this new world with Him. He could have created anything He wanted, of course, by any means He cared to use. But He had already decided that the living world would be physically independent of direct divine intervention, and that life would find its support in the physics and chemistry that He was careful to create.

How should He bring His creatures into existence? The storytellers of intelligent design assert that He could have done it just one way—by assembling every element, every interlocking bit and piece of tissue, cell, organ, and genetic code in single flashes of creative intensity. Since His creatures would inhabit a planet in which the flow of nutrients, the complex cycles of carbon, nitrogen, and oxygen, and the balanced intricacy of food chains and nutrient webs would be required to sustain life, all of these would also have to be assembled with the same care, involvement, and detail given to each of the billions of organisms spread across the face of the planet.

Curiously, once set in motion, the ordinary operations of physics and chemistry would be more than sufficient to adjust these networks and cycles to the changing challenges of life on the planet; and just as curiously, new individuals would find their origins in the ordinary processes of cellular life, dependent in every respect upon the very substance of matter. In other words, God would use material processes to bring their individual lives into existence, although the very same processes were not sufficient to bring their species into existence. Did the world of the Creator require Him to fashion such a stunning discontinuity between the present and the past? I think not.

Having decided to base life on the substance of matter and its fine-tuned properties, a Creator who had already figured out how to fashion the beauty of order without the hobble of determinism could easily have saved His greatest miracle for last. Having chosen to base the lives of His creatures on the properties of matter, why not draw the origins of His creatures from exactly the same source? God's wish for consistency in His relations with the natural world would have made this a perfect choice. As His great creation burst forth from the singularity of its origin, His laws would have set within it the seeds of galaxies, stars, and planets, the potential for life, the inevitability of change, and the confidence of emerging intelligence.

When creatures evolved who were ready to know Him, beings worthy of souls, His work would have reached a pinnacle of power and subtlety. That is the point, the religious person might say, where God decided to reveal Himself to us. And that revelation, in all traditions, is understood as persuasive without being forceful, compelling and not coercive. The Western God stands back from His creation, not to absent Himself, not to abandon His creatures, but to allow His people true freedom. A God who hovers, in all His visible power and majesty, over every step taken by mere mortals never allows them the independence that true love, true goodness, and true obedience requires.

For our freedom in this world to be genuine, we must have the capacity to choose good or evil, and we must be allowed to face the consequences of our own actions. No God could have created individuals who were free to sin but *never* chose to do so. As Barbour points out, "Virtues come into being only in the moral struggle of real decisions, not ready-made by divine fiat."[28] Having lived through much of a century that has seen unspeakable evil visited upon mankind, I think it clear that no person could maintain their faith in God's goodness if they believed He was the reason for such bloodshed and horror. He is not. Our sins are made on earth, not heaven. If we choose to trace any of them to God, we will find at their sources only the value He places on human freedom.

The ultimate purpose of the work of this God may never be understood by the mind of man. Perhaps it was, as the Baltimore Catechism told me long ago, that God wanted to be known, loved, and served. If that is true, He did so by devising a universe that would make knowledge, love, and service meaningful. Seen in this way, evolution was much more than an indirect pathway to get to you and me. By choosing evolution as His way to fashion the living world, He emphasized our material nature and our unity with other forms of life. He made the world today contingent upon the events of the past. He made our choices matter, our actions genuine, our lives important. In the final analysis, He used evolution as the tool to set us free.

The Book of Beginnings

Evolution has traditionally been seen as a problem for religion, a troubling boulder in the pathway to God around which believers must tiptoe. Nearly all of the resistance to evolution derives from the reluctance of religion to make this adjustment. Believers who proclaim their acceptance of evolution are presumed to have found a way to compromise or reconfigure the key elements of their faith in order to sidestep the problems presented by scientific materialism. However they have done this, the presumption is that they have found a way to bend their understanding of God to fit the harsh and unyielding facts of science.

Nonbelievers, by the same token, feel absolved from such adjustments. Evolution, they maintain, is proof of the nonexistence of any traditional Deity, testament to an absence of purpose to the universe, and proud foundation to a philosophy of absolute materialism. Richard Dawkins speaks for many when he writes: "The more you understand the significance of evolution, the more you are pushed away from the agnostic position and towards atheism."[29]

In the late nineteenth century, the German naturalist Ernst Haeckel was more than willing to step outside of biology to make this point explicit:

> The cell consists of matter called protoplasm, composed chiefly of carbon with an admixture of hydrogen, nitrogen, and sulfur. These component parts properly united produce the soul and body of the animated world, and suitably nursed became man. With this single argument the mystery of the universe is explained, the Deity annulled, and a new era of infinite knowledge ushered in.[30]

Haeckel's prequantum fantasy of infinite knowledge, a product of nineteenth century rationalism, dissolved a century ago, but the emotional certainty of his absolute materialism remains.

Haeckel's archaic certainty no doubt owes more to the Book of Genesis than to any book of biochemistry. There is simply no theological reason to argue that life cannot have a material basis. There is, however, an abundance of scientific evidence showing that Genesis could not possibly be a step-by-step, factual account of actual events in natural history. Haeckel, like absolute materialists today, sees this as a fatal flaw. Since Christianity and Judiasm base their beliefs on the Bible, the fact that the opening pages of the good book are scientifically incorrect means that the whole thing is rubbish. Biblical literalists, holding the other side of the absolutist coin, use the same conflict to conclude that evolution is rubbish. Both sides are wrong.

The irony is that even the most extreme critics of evolution are not biblical literalists, including three of the anti-evolutionists—Berlinski, Johnson, and Behe—whom we met in earlier chapters.[31] Berlinski claims not to be a believer. Johnson and Behe, both professing Christians, are not biblical literalists. Neither, for that matter, were the early fathers of the Church, including Saint Basil and most especially Saint Augustine. For Christians, Saint Augustine's understanding of Genesis is critical. His work, *The Literal Meaning of Genesis,* was a detailed attempt to explore every aspect of the Bible's first book in the prescientific scholarly traditions of early Christianity. One may dispute Augustine, but one could hardly call him either anti-Christian or an apologist for Darwin. Keeping that in mind, Augustine's reading of Genesis is tough going for anyone who regards the words of the Bible as distinctly hostile to an evolutionary view of life. Key to this, as physics professor Howard Van Till describes it, was Augustine's conception of the character of the world:

> The universe was brought into being in a less than fully formed state but endowed with the capacities to transform itself, in conformity with God's will, from unformed matter into a marvelous array of structures and lifeforms. In other words, Augustine envisioned a Creation that was, from the instant of its inception, characterized by functional integrity.[32]

Augustine was not an evolutionist. Nonetheless, he saw creation as a continuing and unfolding process in which the commands of the Creator were fulfilled progressively, not instantaneously. And more to the point, he was adamant that even the "literal" meaning of Genesis must not stand in contradiction to the kind of knowledge that today we would call "scientific."

> Even a non-Christian knows something about the earth, the heavens, and the other elements of this world, about the motion and orbit of the stars and even their size and relative positions, about the predictable eclipses of the sun and moon, the cycles of the years and the seasons, about the kinds of animals, shrubs, stones, and so forth, and this knowledge he holds to as being certain from reason and experience. Now it is a disgraceful and dangerous thing for an infidel to hear a Christian, presumably giving the meaning of Holy Scripture, talking nonsense on

> these topics; and we should take all means to prevent such an embarassing situation, in which people show up vast ignorance in a Christian and laugh it to scorn.[33]

Sacred scripture, in Augustine's words, "has been written to nourish our souls," not to present us with a scientific description of the world. He goes on to scold those who put forward interpretations of Genesis that any scientifically knowledgeable non-Christian would recognize as nonsense:

> Reckless and incompetent expounders of Holy Scripture bring untold trouble and sorrow on their wiser brethren when they are caught in one of their mischievous false opinions and are taken to task by those who are not bound by the authority of our sacred books.[34]

How might a Christian approach those who claim that the Genesis narrative teaches a natural history that *must* contradict the evolutionary account? Perhaps by reminding them, as Augustine did, that great damage is done to Christian belief by asserting in the face of scientific fact that *their* contradictory readings of Scripture must be true.

How, then, is a Christian to read Genesis in light of evolution and the knowledge it provides from reason and experience? I suggest that the answer is in Genesis itself, and it fully supports Augustine's view. As more than one modern reader has noted, Genesis 1 and Genesis 2 present creation narratives differing dramatically in their essential details. For example, Chapter 1 describes the simultaneous creation of man and woman in the same verse (27):

> So God created man in his own image, in the image of God created he him; male and female created he them.

In Chapter 1 we don't learn exactly *how* the creation was accomplished, and we don't learn the actual substance from which male and female were formed. But in Chapter 2 the story is told again, and in verse 7 we read something quite different:

> And the Lord God formed man of the dust of the ground, and breathed into his nostrils the breath of life; and man became a living soul.

This time man is made first; and before woman is created, a great deal happens. God places the man in Eden, causes trees to grow forth from the ground, produces rivers to water the trees, tells the man to avoid the fruit of the tree of knowledge, and finally requires the man to give a name to *every* living creature. Only much later (Genesis 2:22), when Adam goes to sleep, is Eve, the first woman, created. The obvious conflict between these two accounts requires even a biblical literalist to apply interpretation to fix upon the meaning of each passage. I make this point not to criticize any particular reading of Genesis, but rather to observe that *any* reading of that book requires interpretation and judgment.[35]

So, what does this mean for any believer, Muslim, Jew, or Christian, who sees scripture as the word of God? To any biochemist, even an evolutionary biochemist, the notion that human life was formed from the dust of the earth is not only poetic, but scientifically accurate to an astonishing degree. An extreme literalist—of the sort abjured by Augustine—might use Genesis 2:7 to argue that the elemental composition of the human body should *match* that of ordinary dust.[36] A broader and more sensible reading would tell us simply that the materials of the human body were taken from the earth itself, which of course is true. To understand Genesis, to find the greater truth, I would argue, all one has to do is to apply the more sensible reading throughout.

As a scientist, I know very well that the earth is billions of years old and that the appearance of living organisms was not sudden, but gradual. As a Christian, I believe that Genesis is a true account of the way in which God's relationship with the world was formed. And as a human being, I find value in both descriptions. In order to reveal Himself to a desert tribe six thousand years ago, a Creator could hardly have lectured them about DNA and RNA, about gene duplication and allopatric speciation. Instead, knowing exactly what they would understand, He spoke to them in the direct and lyrical language of Genesis. He indicated His will that the waters "bring forth abundantly" creatures that swim, that the earth "bring forth" all manner of plants and animals, and spoke freely of having "created" and "made" many other species. His will to create, of course, was reflected in the construction of matter itself, from the laws of chemistry to the gravitational constant, and made the evolution of life in the universe a certainty. By any standard, God's work in creating the universe amounts—literally—to a command that the earth and its waters bring forth life. The words of Genesis do not amount to natural history, but to a believer they do reflect a true and accurate spiritual history of God and His creation.

Augustine warned early Christians against making scientific extrapolations from the powerful imagery of Genesis, He spoke of the "light" created by God as being spiritual, not physical, and He made it clear that the "days" encompassed in Genesis 1 were not to be understood as ordinary twenty-four-hour days.[37] Today we can see, with Augustine's guidance, that evolution does not deny the biblical account, but rather completes it.

For far too long, the critics of religion have used Genesis as a convenient punching bag. Their seconds have been creationists and biblical literalists eager to abandon scientific fact and read Genesis as a history text rather than a narrative "written to nourish our souls." But Genesis, properly understood, does not lend itself to such abuse. The God of the Bible, even the God of Genesis, is a Deity fully consistent with what we know of the scientific reality of the modern world.

All too often, religious people find themselves trying to find a way of apologizing religious belief into scientific fact. As a scientist and a Christian, I do not believe that such apology is necessary. What we have learned from science explains, for the first time in human history, how God could have solved the overpowering logical problems of His divine nature by creating for us a distinctive world of meaning and substance.

To the nonbeliever, that world exists and operates entirely in the absence of God. The self-sufficiency of nature, exemplified most notably by evolution, implies an autonomy that requires neither explanation nor justification in divine terms. Scientifically, that self-sufficiency can be identified, tested, theorized, and explored, but its meaning and the reason for its existence cannot be explained or even addressed by science. This—and not any departure from the discipline of science—is what distinguishes a believer from a nonbeliever. To a believer, the world makes sense, human actions have a certain value, and there is a moral order to the universe.

The irony is that only those who embrace the scientific reality of evolution are adequately prepared to give God the credit and the power He truly deserves. By recognizing the continuing force of evolution, a religious person acknowledges that God is every bit as creative in the present as He was in the past. That—and not a rejection of any of the core ideas of evolution—is why I am a believer. A strong and self-confident religious belief cannot forever pin its hopes on the desperate supposition that an entire branch of science is dramatically wrong, thereby to teeter always on the brink of logical destruction. To be sure, genuine faith requires from its adherents a trust in God, but it also demands a confidence in the power of the human mind to investigate, explore, and understand the evolving nature of God's world.

9

FINDING DARWIN'S GOD

The great hall of the Hynes Convention Center in Boston looks nothing like a church. So what was I doing there, smiling in the midst of an audience of scientists, shaking my head and laughing to myself as I remembered another talk, given long ago, inside a church to an audience of children?

Without warning, I had experienced one of those moments in the present that unexpectedly connects with the scattered recollections of our past. Psychologists tell us that such things happen all the time. Five thousand days of childhood are filed, not in chronological order, but as bits and pieces linked by words, or sounds, or even smells that cause us to retrieve them for no apparent reason when something refreshes our memory. And just like that, a few words in a symposium on developmental biology had brought me back to the day before my First Communion. I was eight years old, sitting with the boys on the right side of our little church (the girls sat on the left) and Father Murphy, our pastor, was speaking.

Trying to put the finishing touches on our year of preparation for the sacrament, Father was trying to impress us with the reality of God's power in the world. He pointed to the altar railing, its polished marble gleaming in sunlight, and firmly assured us that God Himself had fashioned it. "Yeah, right," whispered the kid next to me. Obviously worried that there might be the son or daughter of a stonecutter in the crowd, the good Father retreated a bit. "Now, he didn't carve the railing or bring it here or cement it in place . . . but God Himself *made* the marble, long ago, and left it for someone to find and make into part of our church."

I don't know if our pastor sensed that his description of *God as craftsman* was meeting a certain tide of skepticism, but no matter. He had another trick up his sleeve, a can't-miss, sure-thing argument that no doubt had never failed him. He walked over to the altar and picked a flower from the vase.

"Look at the beauty of a flower. The Bible tells us that even Solomon in all his glory was never arrayed as one of these. And do you know what? Not a single person in the world can tell us what makes a flower bloom. All those scientists in their laboratories, the ones who can split the atom and build jet planes and televisions, well, not one of them can tell you how a plant makes flowers." And why should they be able to? "Flowers, just like you, are the work of God."

I was impressed. No one argued, no one wisecracked. We filed out of the church like good little boys and girls, ready for our First Communion the next day. And I never thought of it again, until this symposium on developmental biology. Sandwiched between two speakers working on more fashionable topics in animal development was Elliot M. Meyerowitz, a plant scientist at Caltech. A few of my colleagues, uninterested in any research dealing with plants, got up to stretch their legs before the final talk, but I sat there with an ear-to-ear grin on my face. I jotted notes furiously, I sketched the diagrams he projected on the screen, and wrote additional speculations of my own in the margins. Meyerowitz, you see, had explained how plants make flowers.

The four principal parts of a flower—sepals, petals, stamens, and pistils—are actually modified leaves. This is one of the reasons why plants can produce reproductive cells just about anywhere, while animals are limited to a very specific set of reproductive organs. Your little finger isn't going to start shedding reproductive cells anytime soon. But in springtime, the tip of any branch on an apple tree may very well blossom and begin scattering pollen. Plants can produce new flowers anywhere they can grow new leaves. Somehow, however, the plant must find a way to "tell" an ordinary cluster of leaves that they should develop into floral parts. That's where Meyerowitz's lab came in.

Several years of patient genetic study had isolated a set of mutants that could form only two or three of the four parts. By crossing the various mutants, his team was able to identify four genes that have to be turned on or off in a specific pattern to produce a normal flower. Each of these genes, in turn, sets off a series of signals that "tell" the cells of a brand-new bud to develop as sepals or petals rather than ordinary leaves. The details are remarkable, and the interactions between the genes are fascinating. To me, sitting in the crowd thirty-seven years *after* my First Communion, the scientific details were just the icing on the cake. The real message was, "Father Murphy, you were wrong." God doesn't make a flower. The floral induction genes do.

Our pastor had made a mistake. Knowing all too well that there was much in nature that science had not mastered, he sought proof for the existence of God in one of those unsolved mysteries. "Explain this!" was the challenge he might have seen himself throwing at the feet of science. Hearing silence, he could turn back to his eight-year-old charges and say, "See. There's your proof."

Passing up a trial subscription to *Scientific American* was not Father Murphy's critical mistake. His assumption was that God is best found in territory unknown, in the corners of darkness that have not yet seen the light of understanding. These, as it turns out, are exactly the wrong places to look. And even Father Murphy should have known that.

Knowing that science in 1956 had not yet explained the production of flowers, he accepted this as proof that it *never* would. On that score, it would take three or four decades to prove him wrong; but his mistake went well beyond the bad luck of choosing a problem that would eventually be solved. He embraced the idea that God finds it necessary to cripple nature. In our pastor's view, God would not have fashioned a consistent, self-sufficient material universe that could support the blooming of a daffodil without His direct intervention. We can find God, therefore, by looking for things around us that lack material, scientific explanations.

Searching the Shadows

In his own way, our parish priest had taken a page from the creationists' book. They claim that the existence of life, the appearance of new species, and most especially the origins of mankind have not and cannot be explained by evolution or any other natural process. By undercutting the self-sufficiency of nature, they believe, we can find God (or at least a designer) in the deficiencies of science.

The danger to both arguments is clear. Science, given enough time, has explained things that seemed baffling to the best minds of the past. That leads to the conclusion that natural phenomena will have naturalistic explanations, and suggests that creationists would be well-advised as a matter of strategy to avoid telling scientists what they will never be able to figure out. History is against them.

Before the age of science, one might have argued that the world would never yield its secrets to the feeble powers of the human mind. Neither the towers of heaven nor the depths of the earth were accessible to man, and life in all its forms seemed the greatest mystery of all. As we know, all of that has changed. We have walked on the moon, probed the depths of the skies, and even decoded the secrets of life. The good old days of utter mystery may not be gone, but they are fading fast. And a scientific detective list of solved cases, like it or not, includes evolution.

Science writer John Horgan, in his 1996 book *The End of Science,*[1] made exactly this point in a chapter entitled "The End of Evolutionary Biology." No doubt to the disappointment of creationists, the "end" of his title did not proclaim the collapse of a dying theory, but the finality of Darwinism's ultimate triumph. As Horgan explained, the revisionists, pretenders, and critics of *The Origin* have all come up short: "If any scientific idea has proved its ability to overcome all challengers, it is Darwin's theory of evolution."[2] Horgan's book was controversial among scientists because of its bold suggestion that science is running out of problems to solve. As a researcher, I don't much like that idea, and neither do most of my colleagues. Nonetheless, at the core of his thesis was an observation that met with agreement among most of the scientists I know—namely, that in a general way, we really do understand how nature works. And evolution forms a critical part of that understanding.

Contrary to the views of creationists that evolution is untested dogma, the core ideas of evolutionary biology have been subjected to examination time and time again in the ordinary conduct of experimental and historical biology. As Horgan noted, one of the biggest problems facing those who pursue careers in evolutionary biology is how to make their mark on a discipline in which the fundamental problems have already been solved, and solved brilliantly:

> The discipline of evolutionary biology can be defined to a large degree as the ongoing attempt of Darwin's intellectual descendants to come to terms with his overwhelming influence.[3]

On one level, this is exactly why the creationist critique of evolutionary biology is wrong. Evolution really does explain the very things that its critics say it does not, as we saw in the opening chapters of this book. Claims disputing the antiquity of the earth, the validity of the fossil record, and the sufficiency of evolutionary mechanisms vanish upon close inspection. Michael Behe was correct to point out that Darwinian explanations of biochemical machines are rare, but his arguments require that they be absolutely nonexistent. When a paper explaining the biochemical evolution of the Krebs cycle appeared in the same year as his book, his assertion that no such explanation could *ever* be found was exposed as just so much wishful thinking.

As the fossil record becomes more complete, it sometimes provides the very intermediate forms the nonexistence of which the creationists were willing to predict in writing. To pick just one telling example, many such writers, including Michael Behe and Dean Kenyon,[4] were foolish enough to assert in print that paleontologists would *never* discover transitional fossils linking land mammals to the cetaceans, swimming mammals such as whales and dolphins. Arch-creationist Duane Gish went so far as to say that such intermediate creatures were biologically impossible, and put audiences in stitches with an amateurish cartoon showing a ridiculous creature that was half cow, half whale. Unknown to most of them, in the mid-1980s Phillip Gingerich and his associates at the University of Michigan had unearthed important clues suggesting that this evolutionary transition took place near the Indian subcontinent roughly 50 million years ago. Once science knew where to look, the anti-evolutionists were in trouble.

By 1994, Gingerich and fellow paleontologists, including Hans Thewissen, had found not one, but *three* intermediate species[5] linking land mammals to the archeocetes, the oldest swimming mammals. The midpoint of the series, a marvelous animal called *Ambulocetus natans* (the "swimming whale who walks," figure 9.1), displayed exactly the combination of terrestrial and aquatic adaptations that critics of evolution had called impossible, even in principle.[6]

I have no idea whether or not Duane Gish still uses his silly (and comically incorrect) slide, but to anti-evolutionists, the pattern should be clear. Even their favorite gaps are filling up, and the historical record of evolution becomes more compelling with each passing season. This means that science can answer the challenge to evolution in an obvious way. Show the historical record, provide the data, reveal the mechanism, highlight the convergence of theory and fact.

Asking for Trouble

There is, however, a deeper problem caused by the opponents of evolution, but it is not a problem for science. It is a problem for religion. Like our priest, they have based their search for God on the premise that

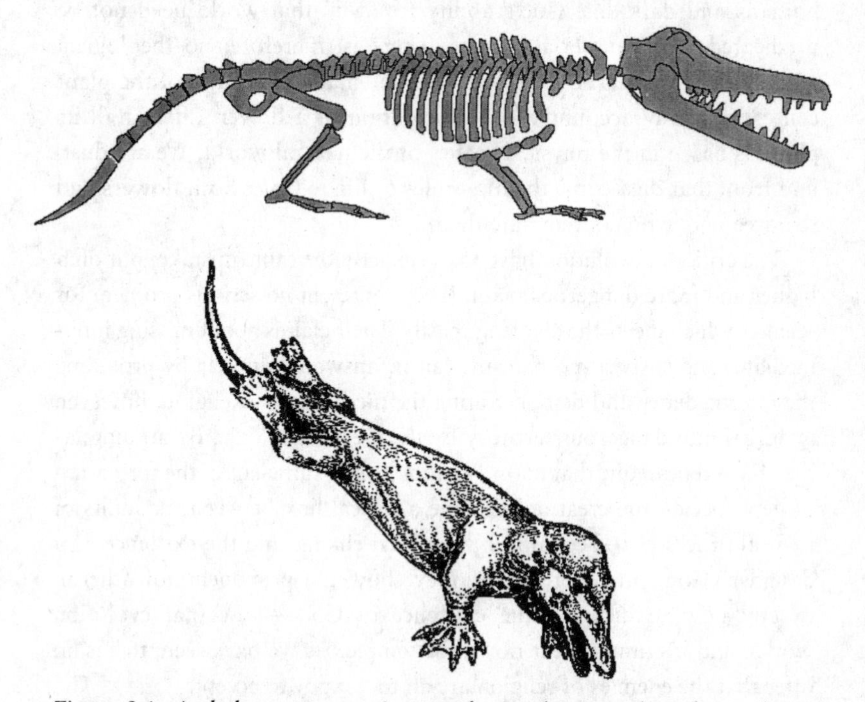

Figure 9.1. *Ambulocetus natans* is one of several intermediate forms documenting the evolution of cetaceans from terrestrial mammals. The skeletal remains of the animal (above) provided Hans Thewissen and his associates with front and hind limbs, establishing that the animal could move easily both on land and in the water.

nature is *not* self-sufficient. By such logic, just as Father Murphy claimed that only God could make a flower, they claim that only God could have made a species. Both assertions support the existence of God *only* so long as they are shown to be true, but serious problems for religion emerge when the assertions are shown to be *false*.

If a *lack* of scientific explanation is proof of God's existence, the counterlogic is unimpeachable: a successful scientific explanation is an argument *against* God. That's why this reasoning, ultimately, is much more dangerous to religion than it is to science. Eliot Meyerowitz's fine work on floral induction suddenly becomes a threat to the divine, even though common sense tells us it should be nothing of the sort.

The reason it doesn't, of course, is because the original premise is flawed. The Western God created a material world that is home to both humans and daffodils. God's ability to act in that world need not be predicated on its material defects. There is, therefore, no theological reason for any believer to assume that the macromolecules of the plant cell cannot fully account for the formation of a flower. Life, in all its glory, is based in the physical reality of the natural world. We are dust, and from that dust come the molecules of life to make both flowers and the dreamers who contemplate them.

The critics of evolution have made exactly the same mistake, but on a higher and more dangerous plane. They represent no serious problem for science, which meets the challenge easily. Their claims about missing intermediates and suspect mechanisms can be answered directly by providing the intermediates and demonstrating the mechanisms. Religion, however, is drawn into dangerous territory by the creationist logic. By arguing, as they have repeatedly, that nature *cannot* be self-sufficient in the formation of new species, the creationists forge a logical link between the limits of natural processes to accomplish biological change and the existence of a designer (God). In other words, they show the proponents of atheism exactly how to disprove the existence of God—show that evolution works, and it's time to tear down the temple. As we have seen, this is an offer that the enemies of religion are all too happy to accept.

Once again, the premise of the argument is flawed. If their God exists, He acts in the world today in concert with natural laws and works His will in the present through the contingent events of human and natural history. All that evolution does is to point out that the workings of natural processes are also sufficient to explain the contingent events of natural history in the past, including the origin and extinction of species. There is neither logical nor theological basis for excluding God's use of natural processes to originate species, ourselves included. There is therefore no reason for believers to draw a line in the sand between God and Darwin. The opponents of evolution have put their money on the wrong horse, and they fail to see that betting so consistently against science is a losing proposition—not for science, but certainly for religion.

As a Christian, I find the flow of their logic particularly depressing. Not only does it teach us to fear the acquisition of knowledge, which might at any time *disprove* belief, but it suggests that God dwells only in the shadows of our understanding. I suggest that if God is real, we should be able to find Him somewhere else—in the bright light of human knowledge, spiritual *and* scientific.

Faith and Reason

Each of the great Western monotheistic traditions sees God as truth, love, and knowledge. Each and every increase in our understanding of the natural world should be a step towards God, and not, as many people assume, a step away. If faith and reason are both gifts from God, then they should play complementary, not conflicting, roles in our struggle to understand the world around us.

Understanding evolution and its description of the processes that gave rise to the modern world is an important part of knowing and appreciating God. As a scientist and as a Christian, that is exactly what I believe. True knowledge comes only from a combination of faith and reason.

A nonbeliever, of course, puts his trust in science and finds no value in faith. And I certainly agree that science allows believer and nonbeliever alike to investigate the natural world through a common lens of observation, experiment, and theory. The ability of science to transcend cultural, political, and even religious differences is part of its genius, part of its value as a way of knowing. What science cannot do is to assign either meaning or purpose to the world it explores. This leads some to conclude that the world as seen by science is devoid of meaning and absent of purpose. It is not. What it does mean is that our human tendencies to assign meaning and value must transcend science, and ultimately must come from outside of it. The science that results, I would suggest, is enriched and informed from its contact with the values and principles of faith. The God of Abraham does not tell us which proteins control the cell cycle. But He does give us a reason to care, a reason to cherish that understanding, and above all a reason to prefer the light of knowledge to the darkness of ignorance.

As more than one scientist has said, the truly remarkable thing about the world is that it actually does make sense. The parts fit, the molecules interact, the darn thing works. To people of faith, what evolution says is that nature is complete. God fashioned a material world in which truly free, truly independent beings could evolve. He got it right the very first time.

In obvious ways, the various objections to evolution take a narrow view of the capabilities of life—but they take an even narrower view of the capabilities of the Creator. They hobble His genius by demanding that the material of His creation ought not to be capable of generating complexity. They demean the breadth of His vision by ridiculing the notion that the materials of His world could have evolved into beings with intelligence and self-awareness. And they compel Him to descend from heaven onto the factory floor by conscripting His labor into the design of each detail of each organism that graces the surface of our living planet.

Sadly, none of this is necessary. If we can accept that the day-to-day actions of living organisms are direct consequences of the molecules that make them up, why should it be any more difficult to see that similar principles are behind the *evolution* of those organisms. If the Creator uses physics and chemistry to *run* the universe of life, why wouldn't He have used physics and chemistry to *produce* it, too?

The discovery that naturalistic explanations can account for the workings of living things neither confirms nor denies the idea that a Creator is responsible for them. To believers, however, it does signify something important. It shows that their God created not a creaky little machine requiring constant and visible attention, but a true, genuine, independent world in which our existence is the product of material forces. Those who *choose* to reject God already know (and so do we) that they need not live in fear of His hand reaching into the sandbox to check our childish actions. God loves us, but He is perfectly willing to allow us to make our own mistakes, commit our own sins, make war on ourselves, and ravage the planet that is our home.

To some, the murderous reality of human nature is proof that God is absent or dead. The same reasoning would find God missing from the unpredictable fits and turns of an evolutionary tree. But the truth is deeper. In each case, a Deity determined to establish a world that was truly independent of His whims, a world in which intelligent creatures would face authentic choices between good and evil, would have to fashion a distinct, material reality and then let His creation run. Neither the self-sufficiency of nature nor the reality of evil in the world mean God is absent. To a religious person, both signify something quite different—the strength of God's love and the reality of our freedom as His creatures.

The Weapons of Disbelief

At least a few of my learned colleagues in evolutionary biology are fond of saying that evolution does have something profound to tell us about the meaning of life—which is that it does not have one. Douglas Futuyma's comments along these lines are typical:

> Some shrink from the conclusion that the human species was not designed, has no purpose, and is the product of mere mechanical mechanisms—but this seems to be the message of evolution.[7]

As heartfelt as such bleak pronouncements may be, they occupy a curious position on the landscape of scientific logic. The strength of science, we are told, is the impartial objectivity it applies to nature, even to human nature. Questions about good and evil, about the meaning and purpose of existence, the sorts of things that have busied philosophers since ancient times, have no place in science, because they cannot be addressed by the scientific method. By what logic, then, do so many invoke science when they presume to lecture on the pointlessness of existence? Something is not quite right. Apparently it is fine to take a long, hard look at the world and assume scientific authority to say that life has no meaning, but I suspect I would be accused of anti-scientific heresy if I were to do the converse, and claim that on the basis of science I had detected a purpose to existence.

The concept of purpose, my colleagues would be quick to remind me, stands outside of science. I agree. But if it does, then so does its exact opposite, that the human species has no purpose, which is the claim made by Futuyma.

I will not pretend to know why so many scientists feel free to invoke the authority of their discipline when they assert a lack of meaning and value to the world. Perhaps it is symptomatic of a culture of atheism, maybe it is a willingness to imprint one's own lack of spiritual values onto science, or perhaps it is a delayed reaction to perceived hostility from religion itself. After dealing with so many religiously inspired challenges to scientific inquiry, maybe it's just human nature to strike back with first rhetorical weapon at hand, the special authority of science.

A religious person should be able to respect such statements, however hostile, as honest disagreements about the nature of things—but he need not and should not accept them as the necessary conclusions of science itself.

For several hundred years, anti-theists have gotten off easy by allowing Genesis to do the hard work for them. An extreme literal reading of Genesis implies that the earth is only a few thousand years old, that all organisms were simultaneously created, and a single worldwide flood accounts for nearly all extinctions and for the whole of the geological record. By the end of the eighteenth century, it was already clear to natural scientists that the actual history of planet earth was quite different from this, and most were quick to reject the scientific authority of Genesis as a result.

Unfortunately, as we saw in Chapter 3, the lure of this simplistic reading of Genesis is still with us. As I have tried to argue, the most effective weapons used by the agents of disbelief are those that believers have handed them willingly. By insisting that Genesis is a scientific document—not a spiritual one—they have made it almost too easy to do exactly what Augustine feared: "to show up vast ignorance in a Christian and laugh it to scorn."

This need not be the case. Even a literal reading of the good book does not require the believer to subscribe to the bogus natural history so easily demolished by scientific fact. As we saw in Chapter 8, our current scientific view of nature is remarkably consonant with the existence of the God we know from Western tradition. Nonbelievers may wish to challenge that assertion, to find other, nonspiritual explanations for the nature of physical reality, the origins of the universe, and the meaning of good and evil. People of faith should look forward to dialogue with them, but they should also, at the very least, demand that nonbelievers carry the heavy logical burden of their absolute materialism all the way to its conclusion, and not provide them with the absurdly easy shortcut of Genesis *as science*.

Having said that, I will add my own observation that even if nonbelievers are denied the easy, first-round knockout afforded by Genesis, most think they still have a winning strategy in reserve—the issue of historic inevitability.

As a species, we like to see ourselves as the best and brightest. We are the intended, special, primary creatures of creation. We sit at the apex of the evolutionary tree, the ultimate products of nature, self-proclaimed and self-aware. We like to think that evolution's goal was to produce *us*.

In a purely biological sense, this comforting view of our own position in nature is false, a product of self-inflating distortion induced by the imperfect mirrors we hold up to life. Yes, we are objectively among the most complex of animals, but not in every sense. Among the systems of the body, we are the hands-down winners for physiological complexity in just one place—the nervous system—and even there a nonprimate, the dolphin, can lay down a claim that rivals our own.

More to the point, any accurate assessment of the evolutionary process shows that the notion of one form of life being more highly evolved than another is incorrect. Every organism, every cell that lives today is the descendant of a long line of winners, of ancestors who found successful evolutionary strategies time and time again, and therefore lived to tell about it—or at least to reproduce. The bacterium perched on the lip of my coffee cup has been through just as much evolution as I have. I've got the advantage of size and consciousness, which matter when I get to write about evolution, but the bacterium has the advantage of numbers, of flexibility, and most especially, of reproductive speed. That single bacterium, given the right conditions, could literally fill the world with its descendants in a matter of mere days. No human, no vertebrate, no animal could boast of anything remotely as impressive.

What evolution tells us is that life spreads out along endless branching pathways from any starting point. One of those tiny branches eventually led to us. We think it remarkable and wonder how it could have happened, but any fair assessment of the tree of life shows that our tiny branch is crowded into insignificance by those that bolted off in a thousand different directions. Our species, *Homo sapiens,* has not triumphed in the evolutionary struggle any more than has a squirrel, a dandelion, or a mosquito. We are all here, now, and that's what matters. We have all followed different pathways to find ourselves in the present. We are all winners in the game of natural selection. *Current* winners, we should be careful to say.

That, in the minds of many, is exactly the problem.

In a thousand branching pathways, how can we be sure that one of them, historically, unavoidably, would lead for sure to us? Consider this: We mammals now occupy, in most ecosystems, the roles of large, dominant land animals. But for most of their history, mammals were restricted to habitats in which only very small creatures could survive. Why? Because another group of vertebrates dominated the earth—until the cataclysmic impact of a comet or asteroid drove those giants to extinction. As Stephen Jay Gould writes:

> If dinosaurs had not died in this event, they would probably still dominate the domain of large-bodied vertebrates as they had for so long with such conspicuous success, and mammals would be small creatures in the interstices of their world. This situation prevailed for 100 million years; why not for 60 million more? . . . In an entirely literal sense, we owe our existence, as large and reasoning animals, to our lucky stars.[8]

So, what if? What if the comet had missed, and what if our ancestors, not the dinosaurs, turned out to be the ones driven to extinction? Or, to use one of Gould's metaphors, what if we wind the tape of life backwards to the Devonian, and imagine the obliteration of the small tribe of fish known as rhipidistians. If they had vanished without descendants, and with them hope of the first tetrapods, vertebrates might never have struggled onto the land, leaving it, in Gould's words, forever "the unchallenged domain of insects and flowers."[9]

No question about it. Rewind that tape, let it run again, and events might come out differently at every turn. Surely this means that mankind's appearance on this planet was *not* preordained, that we are here not as the products of an inevitable procession of evolutionary success, but as an afterthought, a minor detail, a happenstance in a history that might just as well have left us out. I agree.

What follows from this, for skeptic and true believer alike, is a conclusion the logic of which is rarely challenged—that no God would ever have used such a process to fashion His prize creatures. He couldn't have. Because He couldn't have been sure that leaving the job to evolution would have allowed things to work out the "right" way. If it was God's will to produce us, then by showing that we are the products of evolution, we would rule Him out as our Creator. Therein lies the value or the danger of evolution. Case closed?

Not so fast. The biological account of lucky historical contingencies leading to our own appearance on this planet is surely accurate. What does not follow is that a perceived lack of inevitability translates into something that we should regard as incompatible with a divine will. To do so shows no lack of scientific understanding, but it seriously underestimates God, even as He is understood by the most conventional of Western religions.

When examined closely, the notion that we must find historical inevitability in a process in order to square it with the intent of a Creator makes absolutely no sense. Yes, the explosive diversification of life on this planet was an unpredictable and historically contingent process. So, for that matter, were the rise of Western civilization, the collapse of the Roman Empire, and the winning number in last night's lottery. We do not regard the indeterminate nature of any of these events in *human* history as antithetical to the existence of a Creator, so why should we regard similarly indeterminate events in *natural* history any differently? There is no reason at all. If we can look at the contingent events in the families that produced our individual lives as being consistent with a Creator, then certainly we can do the same for the chain of circumstances that produced our species.

Clearly, many people look at the string of historical contingencies leading to our species as something that diminishes the special nature of humankind. What they fail to appreciate is that the alternative, a strictly determined chain of events in which our emergence was preordained, would require a strictly determinant physical world. In such a place, all events would have predictable outcomes, and the future would be open neither to chance nor independent human action. A world in which we would *always* evolve is also a world in we would *never* be free. Seen this way, I think it is only fair for religious people to view the contingency and improbability of our origins as something deeper. The special nature of the particular history that led to us can make us understand how truly remarkable we are, how rare is the gift of consciousness, how precious is the chance to understand, and to the believer, how great are the gifts and expectations of God's love.

Can we really say that no Creator would have chosen an indeterminate, natural process as His workbench to fashion intelligent beings? Gould argues that if we were to go back to the Cambrian era and start over a second time, the emergence of intelligent life exactly 530 million years later would not be certain. I think he is right, but I also think this is less important than he believes. Is there some reason to expect that the God we know from Western theology had to preordain a timetable for our appearance? After 4.5 billion years, can we be sure He wouldn't have been happy to wait a few million longer? And, to ask the big question, do we have to assume that from the beginning he planned intelligence and consciousness to develop in a bunch of nearly hairless, bipedal, African primates? If another group of animals had evolved to self-awareness, if another creature had shown itself worthy of a soul, can we really say for certain that God would have been less than pleased with His new Eve and Adam? I don't think so.

God's statement in Genesis 1:26, "Let us make man in our image, after our likeness,"[10] is sometimes taken to exclude this possibility. A literal reading of "image" and "likeness" would result in the requirement that God look like us, that He be human in form and appearance, which would mean that the emergence of creatures *exactly* like us had to be preordained. Yet God is also said to be a spirit, a nonmaterial being who reveals Himself in many ways—as a burning bush, a dove, tongues of fire. A nonmaterial being could take any form, even the form of a specific person, but theologians have long maintained that any vision of God as a physical person of a particular age, dress, and appearance is necessarily in error. If we are made in the image of God, it must be in some way that transcends physical appearance.

What the Genesis phrase does imply, in the most conventional theological terms, is that God is *like* us. That He understands us. That we can trust our moral senses and faculties to find what He regards as good and true. The common sense of this phrase even extends to science. Some doubt the validity of the quantum picture of matter, suggesting that the submicroscopic world seems strange only because our brains have evolved to function in the macroscopic, "big" world of everyday events. By such reasoning, we should abandon our attempts to fashion a complete description of nature, because the inherent limits of our animal minds make logic itself untrustworthy.

Genesis 1:26 tells us something very different. We are assured that our efforts to understand nature are valid, because our hearts and minds are fashioned in the likeness of God. Our senses, the ultimate sources of scientific knowledge, are imperfect, but they are not deceitful. Nature does not give up her secrets easily, and our first explanations are not always correct. Nonetheless, if we persist, if we apply the tools of reason and the power of the human mind, the world will yield its secrets; and so it has. The God in whom believers put their trust may surpass our understanding, but the assurance that we are made in His image means that He and His world are accessible to our best efforts. The world actually does make sense.

The vastness of the universe itself gives a hint that this was exactly God's approach. If a Creator were to fashion a world in which the constants of matter and energy made the evolution of life *possible,* then by forming millions of galaxies and billions of stars with planets, he would have made its appearance *certain*. With a sample size of only one, we can hardly look at earth's natural history and be assured that the evolution of intelligence and consciousness is the unavoidable outcome of life here or anywhere else. But given the size of the universe, it is easy to imagine that there may be many such experiments in progress. For all we know, God has revealed Himself to us, according to our many religious traditions, because we were the first of these experiments to be ready; or because we were merely the latest of His many encounters with creation.

Given our late start in exploring the cosmos, no one should pretend to know the answers to such questions, but neither should anyone assume that the great range of evolution's possible outcomes somehow runs contrary to the demands of faith. It may demonstrate instead the Creator's determination to fashion a world in which His creatures' individual choices and actions would be free to affect the future. And it is only in such a world that conscious beings would face true moral choices, and know the certainty of genuine peace made possible by such a Creator.

Amazing Grace

Believers since the earliest times have sought certainty, and still do. The opposite of faith is doubt, and both are part of human character. To resolve their doubts, many people of faith seek a *direct* link to God in nature. They hunger for the indisputable miracle in stone, the unevolvable protein, the unbridgable gap in the fossil record—something that will show them a "Made by Yahweh" stamp inside of the machinery of nature. In principle, it is always possible that we will find such a mark, but scientific history argues that it is very unlikely. Time and time again, science has shown that natural phenomena have naturalistic causes, and that the physical logic of the universe is self-consistent.

We could, if we wished, hold up the origin of life itself as an unexplained mystery, and find in that our proof of God at work. Since neither I nor anyone else can yet present a detailed, step-by-step account of the origin of life from nonliving matter, such an assertion would be safe from challenge—but only for the moment. We already know, from careful experiments, that simple molecules of the sort found on the early earth can combine to form many of the larger, more complex molecules required for life. We also know that under certain conditions, some molecules, including RNA, are capable of self-replication. We know that membrane-like barriers can form spontaneously from simple lipids and fatty acids, and we also know that the ancient earth had enormous supplies of energy, solar, chemical, and geothermal, available to drive all manner of chemical reactions. None of these facts proves that life originated purely of naturalistic causes, and therefore none of them proves that the first cell was not the direct, miraculous, intentional work of a Creator. At least not *yet,* and there's the danger.

My point is that there is no religious reason, none at all, for drawing a line in the sand at the origin of life. The trend of science is to discover and to explain, and it would be foolish to pretend that religious faith must be predicated on the inability of science to cross such a line. Evolution, after all, does not require that life must have originated from naturalistic causes—only that its biological history is driven by the same natural forces we observe every day in the world around us. Remember, once again, that people of faith believe their God is active in the present world, where He works in concert with the naturalism of physics and chemistry. A God who achieves His will in the present by such means can hardly be threatened by the discovery that He might have worked the same way in the past.

I devoted much of the first part of this book to reviewing the evidence that supports evolution, and to exposing the intellectual weaknesses of the many challenges against it. One would like to think that all scientific ideas, including evolution, would rise or fall purely on the basis of the evidence. If that were true, public opinion of evolution would long since have passed from controversy into common sense. That is exactly what has happened within the scientific community, but unfortunately, not among the general public. In the minds of many, evolution remains a dangerous idea. For biology educators, it is a source of never-ending strife.

I believe much of the problem lies with atheists in the scientific community who routinely enlist the material findings of evolutionary biology in support their own philosophical pronouncements. Sometimes, as we have seen, these take the form of stern, seemingly dispassionate pronouncements about the meaninglessness of life. Other times, we are lectured that the contingency of our presence on this planet invalidates any sense of human purpose. And very often we are told that the raw reality of nature strips the trapping of authority from any human system of morality.

In the first chapter of this book I quoted an exceptional passage from George C. Williams's compelling book, *The Pony Fish's Glow.*[11] He described Sarah Hrdy's study of infanticide among one of our primate relatives, noting how from time to time a strong outsider will attack and defeat the single dominant male and take charge of his harem. In Williams's words, each female "accepts the advances of her baby's murderer, and he becomes the father of her next child." Williams then asks each of his readers to come to a theological conclusion on the basis of this natural history lesson: "Do you still think God is good?"

No doubt Williams would have argued that if evolution is both the instrument of God and also the source of murder and rape within the harem, then the God of evolution must be a pretty nasty fellow, certainly not the God we think we know through prayer and worship. In making the same argument today, we could go further than Williams. We might note a 1998 Canadian study[12] comparing the rate of infanticide in *human* families headed by a child's biological father to that of families headed by a stepfather. Fortunately, the absolute rate of infanticide is very small in either case. However, infanticide committed by a stepfather occurs at almost sixty times the rate of killing by a child's biological father. We could note, if we cared to, that Hrdy's observations were explicable in evolutionary terms, because in monkeys such infanticide would increase the proportion of the new harem-master's genes in the next generation. Do the human data mean that stepfathers have a preprogrammed evolutionary instinct to kill their new wives' children? And if it does, can we still think God is good?

I do think God is good, and I believe the stepfather data should be read another way. George C. Williams literally wrote the book on evolutionary adaptation, but like many nonbelievers he fundamentally misunderstands what faith in God implies about the nature of good and evil. To begin with, the harem-murders of Hrdy's monkeys are not moral lessons, but animal behaviors. Evolution, under some circumstances, may indeed favor the development of behaviors in which one individual kills others of its own species. For example, the first, instinctive action of a newly emerged queen bee is to sting to death her pupating sisters before they can emerge to challenge her rule. A true naturalist like Williams should recognize that any pronouncement of such behaviors as good or bad does not come from science. We find the murder of a nursing monkey shocking not in itself, and certainly not in its evolutionary logic, but only by analogy, because it is so easy to compare such actions to our own.

Did evolution give human stepfathers the instinct to kill? I think not. The very low rates at which stepchildren are murdered or harmed support a different generalization; namely, that the vast majority of stepfathers are kind, effective, and loving parents. What evolution has done is to imprint upon biological fathers an incredibly strong instinct against harming their own offspring. The evolutionary basis of this is clear—genes that predispose parents *against* the killing their own children are actually helping copies of those genes to survive into the next generation. In this case, what evolution has done is to produce such a strong aversion to infanticide of one's own child that the killings of another's child, however rare, stand out in statistical contrast.

Can God be good when there is evil, even murder, in the living world? Of course He can. To religious people, God is the embodiment of good. Why, then, does He allow evil to exist? How could a kind and loving God allow a father to murder his child? How could He permit the carnage of war, the terror of natural disaster, the inhuman agony of famine and disease? He allows such things because He has made us material creatures, dependent upon the physical world for our existence. In such a world, the destruction of one form of life comes about as a natural consequence of the existence of another. We are connected to that natural world. We are born in pain, we struggle for our food and drink and shelter, we age, and eventually we die. He allows such things as a consequence of the gift of human freedom. The ability to do good means nothing without the freedom to do evil. In a world of individuals, some will always choose the latter, and their actions form an unfortunate backdrop to which the moral choices of virtue, charity, and honesty stand in contrast.

As creatures fashioned by evolution, we are filled, just as E. O. Wilson says, with instinctive behaviors important to the survival of our genes. And there is no doubt that some of these behaviors, though favored by natural selection, can get us into trouble. Our desires for food, water, reproduction, and status, our willingness to fight and our tendencies to band together into social groups, can all be seen as behaviors that help to ensure evolutionary success of individuals who carry such tendencies in their genes. Sociobiology, which studies the biological basis of social behaviors, tells us there are circumstances in which natural selection will favor cooperative and nurturing instincts, "nice" genes that help us get along together. There are also circumstances that will favor aggressive, self-centered behaviors ranging all the way from friendly competition to outright homicide. Could such Darwinian ruthlessness be part of the plan of a loving God?

Naturally, it could. If life on this planet is distinguished by anything, it is by its durability. Species have come and gone, but each new species, each evolutionary branch, each twig of the tree of life represents part of its continuing triumph. So do we. To survive on this planet, the genes of our ancestors, like those of any other organism, had to produce behaviors that protected, nurtured, defended, and ensured the reproductive successes of the individuals that bore them. It should be no surprise that we carry such passions within us, and it is hardly the fault of Darwinian biology for giving their presence a biological explanation. Indeed, the Bible itself gives ample documentation of such human tendencies, including pride, selfishness, lust, anger, aggression, and murder.

Darwin can hardly be criticized for pinpointing the biological origins of these drives. All too often, in finding the sources of our "original sins," in fixing the reasons why our species displays the tendencies it does, evolution is misconstrued as providing a kind of justification for the worst aspects of human nature. At best, this is a misreading of the scientific lessons of sociobiology. At worst, it is an attempt to misuse biology to abolish any meaningful system of morality. Evolution may explain the existence of our most basic biological drives and desires, but that does not tell us it is always proper to act on them. The fact that evolution provided me with a sense of hunger when my nutritional resources are running low does not justify my clubbing you over the head to swipe your lunch. Evolution explains our biology, but it does not tell us what is good, or right, or moral. For those answers, however informed we may be by biology, we must look somewhere else.

Every one of us, as a human being, is born with a raw core of passion that religion knows all too well. That core is part of our biological humanity just as surely as are the bones of our hands or the blood that travels through our veins. The sweeping process of change we know as evolution shaped every part of us, and made us the creatures we are today. Our faults, our strengths, our weaknesses, and our creative imaginations are part of that biological heritage. For those who lack a sense of the spiritual, the reality of that heritage may be all there is to mankind; but for people of faith, there is something more.

There is no scientific way to describe the spiritual concept of grace, which makes it less than real to an absolute materialist. To a believer, grace is as real as the presence of God Himself. Do Darwin's revelations—the discoveries that locate the sources of human passions in survival mechanisms—contradict the reality of grace? Not in the least. To a believer, grace is a gift from God that enables us to place our lives in their proper context—not by denying our biological heritage, but by using it in His service. To be sure, our fears, our desires, our jealousies provide us with reasons to fail, but they also provide us with the means and the opportunity to succeed. To a believer, God's great gift was to provide us with a means to understand, to master, and to do good using both the strengths and weaknesses of human nature.

Where does science sit in all of this? I would argue that any scientist who believes in God possesses the faith that we were given our unique imaginative powers not only to find God, but also to discover as much of His universe as we could. In other words, to a religious person, science can be a pathway towards God, not away from Him, an additional and sometimes even an amazing source of grace.

VESTIGES

In 1844, an extraordinary book was published, *Vestiges of the Natural History of Creation*. The stated theme of the book was that by close examination of the natural world we could understand the natural history of God's creation, maybe even uncover something of the personality of the Creator Himself. At the time, invoking a theological motive for delving into the works of nature was commonplace, but much of what this book had to say was anything but routine.

Author Robert Chambers, an Edinburgh journalist and amateur naturalist, wove his theology throughout his writing, consistently attributing the wonders of the natural world to the magnificence of its Creator:

> Here science leaves us, but only to conclude, from other grounds, that there is First Cause, to which others are secondary and ministrative, a primitive almighty will, of which these laws are merely the mandates. That great Being, who shall say where is his dwelling-place or what his history! Man pauses breathless at the contemplation of a subject so much above his finite faculties, and only can wonder and adore![13]

Pretty conventional, praise-God-as-nature, nineteenth century stuff. But Chambers was enough of a naturalist to understand that geology had been sending very dangerous messages to biology, and he did made sure that his readers understood what they were. One of these, he noted, was that geology seems "inconsistent with the Mosaic record." But that was the least of his shockers. Chambers presented to the public, straight on, the great reality of the fossil record—that species do not appear simultaneously, and that over time "there is a progress of some kind"[14] to the emergence of species on the earth.

Chambers worried that this observation might result in "a storm of unreasoning indignation,"[15] and one could easily view his frequent invocations of the Almighty as intended solely to blunt the religious concerns of his readers. Nonetheless, he was remarkably straightforward about what the progressive appearance of species implied about the work of the divine:

> In what way was the creation of animated beings effected? The ordinary notion . . . [is] that the Almighty author produced the progenitors of all existing species by some sort of personal or immediate exertion. But how does this notion comport with what we have seen of the gradual advance of species, from the humblest to the highest? . . .
>
> Can we suppose that the august being who brought all these countless worlds into form by the simple establishment of a natural principle flowing from his mind, was to interfere personally and specially on every occasion with a new shell-fish or reptile was to be ushered into existence on *one* of these worlds? Surely, this idea is too ridiculous to be for a moment entertained.[16]

So, if not by divine fiat, how did all those forms appear? Chambers suggested that throughout earth's history, a process of "creation by law" had been taking place that accounted for the gradual appearances of animated forms, following rules and principles laid down by the Almighty. As he put it,

> To a reasonable mind the Divine attributes must appear, not diminished or reduced in any way, by supposing a creation by law, but infinitely exalted.[17]

In Chambers's view, a Creator who could set up a process driven by natural law that would drive continuing creation for millions of years was clearly even more clever than a designer who had to do it all "personally and specially," one species at a time.

Chambers anticipated evolution more than a decade before Darwin. He wasn't the first; he was preceded on that score by Erasmus Darwin (Charles's grandfather), Jean Baptiste Lamarck, and several others. But Chambers put it most directly and most publicly. *Vestiges* sold well and was widely read. We hear little about Chambers today, largely because his book was badly weighted down by scientific errors and misconceptions. Thomas Henry Huxley, soon to be Darwin's champion, called it "pretentious nonsense," and Darwin wrote to a colleague that Chambers's "geology strikes me as bad, and his zoology far worse."

Such attacks from the scientific community were mild compared to the theological hostility engendered by the work. Adam Sedgwick wrote, "From the bottom of my soul, I loathe and detest the *Vestiges*,"[18] and other religious critics were even less charitable. Sensibly, Chambers had seen it coming. His work was published anonymously, and he did not allow his connection to *Vestiges* to be revealed until well after his death in 1871.[19] So thorough was his subterfuge that critics of the work were given to referring to its author as "Mr. Vestiges." Charles Darwin had to deny that he was its anonymous author, and its effect on Darwin's own work was significant. According to Gavin de Beer, "In 1844, the same year as that in which *Vestiges* appeared, Darwin had finished his Essay on evolution by natural selection; but seeing how the ideas in *Vestiges* raised the hackles of scientists and theologians alike, he held his hand and kept his secret to himself."[20]

Darwin would eventually write that *Vestiges* "has done excellent service in this country in calling attention to the subject, in removing prejudice, and in thus preparing the ground for the reception of analogous views."[21] As de Beer would put it, Chambers's book served as a lightning rod, sparing Darwin "from at least some of the thunderbolts which would otherwise have fallen on him and the *Origin of Species*."[22] Darwin's magnificent work endured because not only did it avoid Chambers's many biological errors, but it offered something Chambers could not—a mechanism, natural selection, that could drive the processes of evolutionary change.

Despite the significant scientific weaknesses of Chambers's work, it is impossible not to be moved by the strength and sincerity of his conviction that the study of nature is akin to the worship of God. In our own time, human culture has long since grown too sophisticated to pretend that science could be a calling from God. But I worry that such sophistication is self-deceptive.

In graduate school I once attended a seminar that dealt, somewhat indirectly, with the evolution of an important protein. During the question and answer period, another student had the impertinence to note that the speaker's university was religious in nature, and inquired as to whether her mention of the "evolutionary implications" of her work was proper. She smiled. Obviously, it was not the first time she had faced the question. "If you deny evolution, then the sort of God you have in mind is a bit like a pool player who can sink fifteen balls in a row, but only by taking fifteen separate shots. My God plays the game a little differently. He walks up to the table, takes just one shot, and sinks all the balls. I ask you which pool player, which God, is more worthy of praise and worship?" Robert Chambers would have smiled; I certainly did.

I bring up Chambers's quaint, even archaic, attempts to infer the character of the Creator from the nature of His creation because many biologists today seem to believe they can do exactly the opposite—to use nature to *explain away* the Creator, or more specifically, to explain why we are so willing, so gullible as to believe in Him.

Edward O. Wilson fired a shot along these lines, asserting in *On Human Nature* that the tendency to believe in God could be explained by natural selection. If group cohesion is strengthened by common belief in the supernatural, then natural selection will produce, in the most successful human groups, an unreasoning tendency to commit to the supernatural. Having made this link, Wilson thinks he knows why people believe in God, and he also knows what he wants to do with that knowledge:

> If religion, including the dogmatic secular ideologies, can be systematically analyzed and explained as a product of the brain's evolution, its power as an external source of morality will be gone forever.[23]

That clearly is Wilson's goal—to destroy forever religion's power as a source of morality. If religion is just another behavioral product of evolution, how can it possibly claim special authority—indeed, any authority at all—in questions of morals, ethics, and values?

What Wilson never seems to grasp is that science, being also a product of evolutionary forces shaping the human brain, could be discredited by exactly the same logic. If he can explain away a believer's faith as a product of evolutionary forces, that believer can turn around and do exactly the same thing, attributing Wilson's lack of faith to the pressures and demands of evolution on *his* brain. Genes for curiosity and skepticism, valuable as survival tools in the recent past, may well provide the behavioral platform upon which science is built. Wilson's purported scientific views of life and nature could then be explained away as illusory artifacts of primate evolution, constructed to fit the demands of survival, not the objective reality of existence. That evolution helped to *create* our capacities for both faith and science is undeniable. To maintain that either is thereby invalidated fails the test of logic.

Cognitive scientist Steven Pinker approaches religion in a similar way. To Pinker, the brain is a series of "mental modules," each evolved to "solve specialized problems" in the world around us. The modules are formed and shaped by classic Darwinian forces, with each particular module enduring only if its "assumptions worked well enough in the world of our ancestors."[24] One of these, as Pinker and other scientists have said, is a "God module," which predisposes to belief in the Almighty. The existence of this module explains, according to the MIT psychologist, "why a mind would evolve to find comfort in beliefs it can plainly see are false."[25] Pinker does not believe (a subtle difference from Wilson) that this is a distinct adaptation, but a byproduct of the activities of other modules, including "intuitive psychology, the desire for prestige, and acquiescence to experts."[26] Nonetheless, Pinker likewise explains away religion as an unwelcome holdover from our primitive evolutionary past.

The God module is an intriguing way to explain the *capacity* for belief, but as an effort to delegitimize religion, it turns and fails, like Wilson's, on a logical point. If my evolution-honed brain is predisposed to believe in God, by what evolutionary predispositions is Steven Pinker's brain predisposed to believe in modules? Is there a "module module?" By what feat of logic does Pinker declare that *his* brain has risen far enough above its evolutionary influences that he can step outside the bounds of cognitive biology to find a cause for religion, whereas the brains of believers are still mired in their mystical, evolution-induced fogs?

It is a wonderful strategy to claim that you have found a scientific reason to explain why the other guy believes what he does (*he's* the prisoner of biology, of course), and slickly declare that your own beliefs rise above such impediments. In fact, a believer might well say that *of course* people have an evolutionary predisposition to accept the divine. That's how the Creator knew that it was time to endow them with the souls that they believe, as a matter of faith, serve to complete the physical and spiritual reality of human nature. Pinker's evolutionary forces then become just one more tool in the hands of the Almighty.

WHAT KIND OF WORLD?

When I began this book, I recounted the general sentiment that the displacement of God by Darwinian forces in this century is now almost complete. We no longer explain the specializations of an animal or the multiple levels of a food chain in terms of how they fit into God's will, but seek instead to understand the natural forces that shaped each of them over time. As a result, the way in which we appreciate the natural world has changed forever, almost exactly as Charles Darwin had anticipated:

> When the views advanced by me in this volume . . . or when analogous views on the origin of species are generally admitted, we can dimly foresee that there will be a considerable revolution in natural history. . . .
>
> When we no longer look at an organic being as a savage looks at a ship, as at something wholly beyond his comprehension; when we regard every production of nature as one which has had a history; when we contemplate every complex structure and instinct as the summing up of many contrivances, each useful to the possessor, nearly in the same way as when we look at any great mechanical invention as the summing up of the labour, the experience, the reason, and even the blunders of numerous workmen; when we thus view each organic being, how far more interesting, I speak from experience, will the study of natural history become![27]

Well, the views contained within *The Origin* have now been "generally admitted," and the study of natural history has indeed become, as Darwin understated it, "far more interesting." Together with the other makers of modern scientific reality, Darwin lifted the curtain that allowed us to see the world as it really is. And to any person of faith, this should mean that Charles Darwin ultimately brought us closer to an understanding of God.

Darwin himself clearly worried that he had done exactly the opposite. Desmond and Moore, whose excellent biography of Darwin describes these doubts and fears, concluded a stirring summary of the events surrounding Darwin's death and burial in Westminster Abbey with these lines:

> Darwin's body was enshrined to the greater glory of the new professionals who had snatched it. The burial was their apotheosis, the last rite of a rising secularity. It marked the accession to power of the traders in nature's marketplace, the scientists and their minions in politics and religion. Such men, on the up-and-up, were paying their dues, for Darwin

had naturalized Creation and delivered human nature and human destiny into their hands.

Society would never be the same. The "Devil's Chaplain" has done his work.[28]

Which is it? Did Darwin contribute to the greater glory of God, or did he deliver human nature and destiny into the hands of a professional scientific class, one profoundly hostile to religion? Darwin's own views on the subject are so complex and ambiguous that they offer little help. At one time, he said that "agnostic would be the most correct description of my state of mind";[29] but at another, he wrote that he was overwhelmed by

> the extreme difficulty, or rather the impossibility, of conceiving this immense and wonderful universe, including man with his capacity for looking far backwards and far into futurity, as the result of blind chance or necessity. When thus reflecting I feel compelled to look to a First Cause having an intelligent mind in some degree analogous to that of man; and I deserve to be called a Theist.[30]

Cementing his reputation as a fence-sitter, in *The Origin* Darwin took special care to take neither position, declaring his work religiously neutral:

> I see no good reason why the views given in this volume should shock the religious feelings of any one. It is satisfactory, as showing how transient such impressions are, to remember that the greatest discovery ever made by man, namely, the law of the attraction of gravity, was also attacked by Leibnitz, "as subversive of natural and inferentially revealed, religion." A celebrated author and divine has written to me that "he has gradually learnt to see that is it just as noble a conception of the Deity to believe that He created a few original forms capable of self-development into other and needful forms, as to believe that He required a fresh act of creation to supply the voids caused by the actions of His laws."[31]

Darwin, significantly, presented this expansive "conception of the Deity" only as the idea of another, keeping his own views guarded. His cautious language may have been the result of genuine conviction, or could have been intended, as some biographers have written, to spare his family social embarrassment. No matter. The importance of what Darwin has done rises or falls on its own merits, and not on his personal intentions, hopes, or fears. What matters to us today is whether Darwin's work strengthens or weakens the idea of God, whether it serves to enlarge or to diminish a theistic view of the world.

The conventional wisdom is that, whatever one may think of his science, having Mr. Darwin around certainly hasn't helped religion very much. The general thinking is that religion has been weakened by Darwinism, and has been constrained to modify its view of the Creator in order to twist doctrine into conformance with the demands of evolution. As a result, even if we were generous enough to accept science *and* religion as coequal ways of knowing, Orwellian common sense would tell us that one of these partners is more equal than the other. Much more equal. As Stephen Jay Gould puts it, with obvious delight,

> Now the conclusions of science must be accepted *a priori,* and religious interpretations must be finessed and adjusted to match unimpeachable results from the magisterium of natural knowledge![32]

Science calls the tune, and religion dances to its music.

Even the most fervent atheists will stipulate that one can apologize a theistic vision, with due retrospective care, onto almost any scientific reality. This makes God a pesky and elusive target, hard to pin down and impossible to exclude. Nonetheless, to absolute materialists it also means that the aftermath of Darwin is a diminished, roundabout, apologetic version of belief in which religion must constantly be modified to the demands of the scientific moment.

This sad specter of God, weakened and marginalized, drives the continuing opposition to evolution. This is why the God of the creationists requires, above all else, that evolution be shown not to have functioned in the past and not to be working now. To free religion from the tyranny of Darwinism, their only hope is to require that science show nature to be incomplete, and that key events in the history of life can only be explained as the result of supernatural processes. Put bluntly, the creationists are committed to finding permanent, intractable mystery in nature. To such minds, even the most perfect being we can imagine still wouldn't be perfect enough to have fashioned a creation in which life would originate and evolve on its own. The nature they require science to discover is one that is flawed, static, and forever inadequate.

Science in general, and evolutionary science in particular, give us something quite different. Through them we see a universe that is dynamic, flexible, and logically complete. They present a vision of life that spreads across the planet with endless variety and intricate beauty. They suggest a world in which our material existence is not an impossible illusion propped up by magic, but the genuine article, a world in which things are exactly what they seem, in which we were formed, as the Creator once cared to tell us, from the dust of the earth itself.

It is often said that a Darwinian universe is one in which the random collisions of particles govern all events and therefore the world is without meaning. I disagree. A world without meaning would be one in which a Deity pulled the string of every human puppet, and every material particle as well. In such a world, physical and biological events would be carefully controlled, evil and suffering could be minimized, and the outcome of historical processes strictly regulated. All things would move towards the Creator's clear, distinct, established goals. Those who find discomfort in evolution often say that lack of such certainty in the outcome of Darwin's relentless scheme of natural history shows that it could not be reconciled with their faith. Maybe so. But certainty of outcome means that control and predictability come at the price of independence. By being always in control, the Creator would deny His creatures any real opportunity to know and worship Him. Authentic love requires freedom, not manipulation. Such freedom is best supplied by the open contingency of evolution, and not by strings of divine direction attached to every living creature.

The common view that religion must tiptoe around the findings of evolutionary biology is simply and plainly wrong.

One hundred and fifty years ago it might have been impossible not to couple Darwin with a grim and pointless determinism. I believe this is why Darwin in his later years tried and failed to find God, at least a God consistent with his theories. If organisms were mechanisms, and mechanisms were driven only by the physics and chemistry of nature, then we humans were trapped in a material world in which past and future were interlocked in mindless certainty. In such a world, the only chance for God's action would have been in the construction of organisms themselves. Darwin surely felt he had denied himself that refuge by accounting for the illusion of design. As a result, he may well have felt, despite his unwillingness to admit to a world produced by "blind chance or necessity," that he had ruled out any realistic possibility for God. That his God could never be found.

Things look different today. Darwin's vision has expanded to encompass a new world of biology in which the links from molecule to cell and from cell to organism are becoming clear. Evolution prevails, but it prevails with a richness and subtlety its originator may have found surprising, and in the context of developments in other sciences he could not have anticipated.

We know from astronomy that the universe had a beginning, from physics that the future is both open and unpredictable, from geology and paleontology that the whole of life has been a process of change and transformation. From biology we know that our tissues are not impenetrable reservoirs of vital magic, but a stunning matrix of complex wonders, ultimately explicable in terms of biochemistry and molecular biology. With such knowledge we can see, perhaps for the first time, why a Creator would have allowed our species to be fashioned by the process of evolution.

If he so chose, the God whose presence is taught by most Western religions could have fashioned anything, ourselves included, *ex nihilo*, from his wish alone. In our childhood as a species, that might have been the only way in which we could imagine the fulfillment of His will. But we've grown up, and something remarkable has happened—we have begun to understand the physical basis of life itself. If the persistence of life were beyond the capabilities of matter, if a string of constant miracles were needed for each turn of the cell cycle or each flicker of a cilium, the hand of God would be written directly into every living thing—His presence at the edge of the human sandbox would be unmistakable. Such findings might confirm our faith, but they would also undermine our independence. How could we fairly choose between God and man when the presence and the power of the divine so obviously and so literally controlled our every breath? Our freedom as His creatures requires a little space, some integrity, a consistency and self-sufficiency to the material world.

Accepting evolution is neither more nor less than the result of respecting the reality and consistency of the physical world over time. We are material beings with an independent physical existence, and to fashion such beings, any Creator would have had to produce an independent material universe in which our evolution over time was a contingent possibility. A believer in the divine accepts that God's love and His gifts of freedom are genuine—so genuine that they include the power to choose evil and, if we wish, to freely send ourselves to hell. Not all believers will accept the stark conditions of that bargain, but our freedom to act has to have a physical and biological basis. Evolution and its sister sciences of genetics and molecular biology provide that basis. A biologically static world would leave a Creator's creatures with neither freedom nor the independence required to exercise that freedom. In biological terms, evolution is the only way a Creator could have made us the creatures we are—free beings in a world of authentic and meaningful moral and spiritual choices.

Those who ask from science a final argument, an ultimate proof, an unassailable position from which the issue of God may be decided, will always be disappointed. As a scientist I claim no new proofs, no revolutionary data, no stunning insight into nature that can tip the balance in one direction or another. But I do claim that to a believer, even in the most traditional sense, evolutionary biology is not at all the obstacle we often believe it to be. In many respects, evolution is the key to understanding our relationship with God. God's physical intervention in our lives is not direct. But His care and love are constants, and the strength He gives, while the stuff of miracle, is a miracle of hope, faith, and inspiration.

When I have the privilege of giving a series of lectures on evolutionary biology to my freshman students, I usually conclude with a few remarks about the impact of evolutionary theory on other fields, from economics to politics to religion. And I find a way to make it clear that I do not regard evolution, properly understood, as either anti-religious or anti-spiritual. Most students seem to appreciate those sentiments. I expect that they figure that Professor Miller, a nice guy but probably an agnostic, is trying to find a way to be unequivocal about evolution without offending the university chaplain. There are always a few who find me after class and want to pin me down, Usually they ask me point-blank, "Do you believe in God." And one-on-one, I carefully tell them, "Yes." Puzzled, they ask what kind of God? Over the years, I have struggled to come up with a simple but precise answer to that question. Eventually I found it.

I ask my inquiring students to reread the final chapter of Darwin's

NOTES

Chapter 1

1. C. Darwin, *The Origin of Species*, 6th ed. (London: Oxford University Press, 1872, reprinted 1956), p. xxii.
2. J. Milton, *Paradise Lost.* (New York: Penguin Books, 1968), Book VII, lines 379–384.
3. Ibid.
4. Ibid, Book IV, lines 304–314.
5. Darwin, *The Origin,* p. 53.
6. Ibid, p. 65.
7. Ibid, p. 80.
8. D. C. Dennett, *Darwin's Dangerous Idea* (New York: Simon and Schuster, 1995), p. 21.
9. R. Dawkins, *The Selfish Gene,* new ed. (New York: Oxford University Press, 1989), p. 40.

   Dawkins put it this way: "Another general quality that successful genes will have is a tendency to postpone the death of the survival machines at least until after reproduction. No doubt some of your cousins and great-uncles died in childhood, but not a single one of your ancestors did. Ancestors just don't die young!"
10. R. Dawkins, *The Blind Watchmaker* (New York: W.W. Norton, 1986), p. 6.
11. E. O. Wilson, *On Human Nature* (Cambridge: Harvard University Press, 1978), p. 1.

12. R. Dawkins, *River Out of Eden* (New York: HarperCollins, 1995), pp. 95–96.
13. G. C. Williams, *The Pony Fish's Glow* (New York: HarperCollins, 1997), pp. 156–157.
14. Sociobiology, the study of the biological basis of social behavior, can be used to explain the prevalence of infanticide within the monkey harem. Unlearned, instinctive behaviors are presumed to be programmed by genetics, and among the insects and other invertebrates there is good evidence that many of them are. If behaviors are preprogrammed by genes, it stands to reason that natural selection will favor those genes that produce the "best" behaviors, meaning those that do the best job of favoring the reproductive success of those who practice them. In this particular case, even though a strict genetic basis for the behavior of the langurs has not been shown, sociobiologists argue that evolution can account for the success of an infanticide behavior by the new harem-master. By eliminating all currently nursing infants, the harem-master would seem to ensure that his own genes, including ones that might produce or favor the infanticide behavior, would come to dominate the next generation.

Chapter 2

1. R. C. Jastrow, *God and the Astronomers* (New York: W.W. Norton, 1992), p. 9.
2. D. Berlinski, "The Deniable Darwin," *Commentary* 101 (1996), no. 6 (June 1996).
3. An English friend of mine, unfamiliar with the New Hampshire slogan, saw this on a license plate for the first time and shook his head in disbelief. "These people are crazy," he moaned. "No wonder poor George had to let them go." (George III, of course.)
4. Those exceptions include the minuscule amount of matter added to the earth by meteorite impacts and cosmic dust, minus a tiny amount lost as atoms from the upper atmosphere drift into space.
5. My source for these little bits of solar history is *A Star Called the Sun* by George Gamow (New York: Viking Press, 1964).
6. About two centuries later, Eratosthenes did get it right, and calculated a 4,000-mile radius for the earth, remarkably close to the accepted modern value of 3,963 miles.
7. This value refers to the wavelength of the light in nanometers, the usual scientific unit used to describe the wavelength of monochromatic light. A nanometer is one-billionth of a meter (10^{-9} meter).

8. According to my research on this subject, the Schlitz "pop-top" beer can, released in March of 1962, was the first of the so-called easy-open cans. Its removable top was detachable. The success of the can led to a quick adoption of the pop-top idea by soft drink manufacturers in the following year.
9. Those limits are worth noting. We can be reasonably sure of what we find in the past, but we definitely cannot be sure that we have found everything, no matter how hard we look. Scarcity is one reason. Very few organisms generate enduring evidence of their existence at the massive levels that our civilization piles up its garbage. We should expect a few things to be missing. Geography is another. If we dredged our sediments from the East Coast, we might conclude that the Coors Brewing Company had appeared suddenly in 1983, and had never produced a can with a removable pop-top. Both conclusions would be mistaken. Prior to 1983, Coors was distributed only on the West Coast, and even the most dedicated East Coast beer drinkers were unable to make it part of their refuse on a regular basis. The absence of Coors cans from East Coast sediments, therefore, is a cautionary tale against overinterpretation of a single source of data.
10. This excerpt from Smith's journals was taken from the University of California Museum of Paleontology Web site:
http://www.ucmp.berkeley.edu/history/smith.html
11. Ibid.
12. This passage from Cuvier was quoted in *The Growth of Biological Thought* by E. Mayr (Cambridge, MA: Belknap Press, 1982), p. 369.
13. There is a single modern species of gavial, *Gavialis gengeticus,* which is found in Bangladesh, Burma, Pakistan, and India.
14. T. A. Appel, *The Cuvier-Geoffroy Debate* (New York: Oxford University Press, 1987), pp. 131–132.
15. Passage from Geoffroy quoted in *A Short History of Vertebrate Paleontology* by E. Buffetaut (London: Croom Helm, 1987), p. 69.
16. Quotation from the biography of Etienne Geoffroy Saint-Hilaire at the UC Berkeley Museum of Paleontology Web site:
http://www.ucmp.berkeley.edu/history/hilaire.html
17. Dates given on this chart for the first appearance of major living groups are only approximate, and each reflects a consensus reading of the geological literature. I have used the generally accepted names for each geological period, without being concerned whether the name in question represents an era or a period according to the conventions of geology. I have used familiar, rather than scientific, names for most groups, and have not confined them to any one taxonomic level. I have done this to highlight the sequence in which the major vertebrate classes have appeared, and also to

call attention to the emergence of major living groups, especially the flowering plants, since the Cambrian.

18. M. I. Coates and J. A. Clack, "Fish-like Gills and Breathing in the Earliest Known Tetrapod," *Nature* 352 (1991): 234–236.

19. Critics of evolution have often maintained that the earliest amphibian species were fully terrestrial, air-breathing animals. The gills possessed by modern amphibians are external, and quite distinct from the internal gills of fish. The gills of *Acanthostega gunnari,* however, are entirely fish-like, and were supported by the same bones that support the gill arches in primitive fish. Other clues in the skeleton suggest that *Acanthostega gunnari* possessed lungs, and probably could use both gills and air breathing in appropriate circumstances. Coates and Clack concluded their paper with the suggestion that "unique tetrapod characteristics such as limbs with digits evolved first for use in water rather than for walking on land." This suggestion has proven to be remarkably accurate. In 1997, a lobe-finned fish fossil was discovered in which the first traces of digits were preserved, further confirming the validity of the fish-amphibian evolutionary transition. This discovery was reported in "Fish with Fingers?" by E. B. Daeschler & N. Shubin in the journal *Nature* 391 (1998): p. 133.

20. A. D. Desmond and J. Moore, *Darwin: The Life of a Tormented Evolutionist* (New York: Warner Books, 1991), pp. 209–211.

21. Ibid, p. 205, 209–211.

Desmond and Moore make this point in their biography of Darwin: "This was the first fossil Owen diagnosed, and his conclusion was surprising. It belonged to a huge rodent, a hippo-sized capybara relative, which Owen called *Toxodon*. . . . The fossils 'are turning out great treasures,' Charles boasted to Caroline. Rhino-sized rodents! 'What famous cats they ought to have had in those days!' "

As it turns out, Desmond and Moore (and possibly Owen) were mistaken. Capybara is a true rodent, but *Toxodon* belongs to an entirely extinct South American group known as the "notoungulates." *Toxodon* is thought not to be closely related to the capybara, although a number of very large fossil rodents have been discovered on the continent that likely are.

22. Darwin was clearly impressed with the fact that no distinction can truly be made between species and subspecies. On page 53 of *The Origin* he wrote: "Clearly no line of demarcation has as yet been drawn between species and sub-species—that is, the forms which in the opinion of some naturalists come very near to, but do not quite arrive at the rank of species; or again, between subspecies and well-marked varieties, or between lesser varieties and individual differences. These differences blend into each other in an

insensible series; and a series that impresses the mind with the idea of an actual passage."

23. Darwin, *The Origin*, p. 135.
24. T. M. Cronin and C. E. Schneider, "Climatic Influences on Species: Evidence from the Fossil Record," *Trends in Evolutionary Biology and Ecology* 5 (1990): 275–279.
25. This drawing is based on a figure included in Dean Faulk's "Hominid Brain Evolution: Looks can be Deceiving," in *Science* 280 (1998): 1714. Faulk's article described some new—and presumably more accurate—techniques to measure the cranial capacities of key prehuman fossils, and it raised the possibility that much of the data shown in the figure might have to be reevaluated. Although this is doubtless the case, a systematic reevaluation of exact cranial volumes is unlikely to change the pattern seen here—a smooth transition from prehuman (*Australopithecine*) to human (*Homo*) forms.
26. The technical term for the movement of water across a semipermeable membrane, like a bacterial cell membrane, is "osmosis." In this case, it is entirely accurate to say that penicillin kills by osmosis, causing bacteria to make weakened cell walls that cannot stand up to osmotic pressure.
27. These quotations are taken from a news report written by F. Flam, *Science* 265 (1994): 1032.
28. W.P.C. Stemmer, "Rapid Evolution of a Protein in Vitro by DNA Shuffling." *Nature* 340 (1994): 389–391.
29. A. Roth and R. R. Breaker, "An Amino Acid as a Cofactor for a Catalytic Polynucleotide," *Proceedings of the National Academy of Sciences (USA)* 95 (1998): 6027–6031.
30. This was brought home forcefully to me in my first years of college teaching when I cartooned a membrane protein responsible for ion transport as a pump. A pretty good analogy, and one widely used by researchers in the field. To my dismay, when asked to sketch the types of actual structures that might account for the actions of such a protein, several of my students took the pump analogy so literally that they postulated a protein that had the actual parts of a microscopic pump, right down to a rotating piston and a valve system.

Chapter 3

1. From the PBS transcript of a Dec. 4, 1997 *Firing Line* debate entitled, "Resolved: The Evolutionists Should Acknowledge Creation." Buckley spoke in favor of this resolution and I was one of several evolutionists speaking against it.

2. Many attempts have been made to attach a specific date to the creation events described in the first chapter of the Book of Genesis. The most famous of these was the chronology developed in 1654 by the Irish archbishop James Ussher in his *Annales Veteris et Nove Testamenti*. Archbishop Ussher's scholarship fixed the instant of creation at 9 A.M. on October 26, 4004 B.C. Not all young-earth creationists accept the precise Ussher chronology, but all are in general agreement that the earth is less than 10,000 years old, much too young to have allowed evolution time to work.
3. J. C. Whitcomb and H. M. Morris, *The Genesis Flood* (Phillipsburg, NJ: Presbyterian and Reformed Publishing, 1961).
4. K. Chin et al., "A King-sized Theropod Coprolite," *Nature* 393 (1998): pp. 680–682.
5. Several excellent books have dealt with the many fallacies of flood geology. They include:

   D. J. Futumya, *Science on Trial* (New York: Pantheon, 1983).

   L. R. Godfrey, ed., *Scientists Confront Creationism* (New York: W. W. Norton, 1983).

   P. Kitcher, *Abusing Science* (Cambridge, MA: MIT Press, 1984).

   A. Montagu, ed., *Science and Creationism* (New York: Oxford University Press, 1984).

   A. N. Strahler, *Science and Earth History* (Buffalo, NY: Prometheus Books, 1987).
6. H. Morris, *Scientific Creationism* (San Diego: Creation Life Publishers, 1974), p. 136.
7. H. Morris, *The Remarkable Birth of Planet Earth* (San Diego: Creation Life Publishers, 1972), p. 92.
8. Here's how this calculation works: the deepest parts of the ocean are about 5 miles (roughly 25,000 feet). If the tide rises 1 foot in three hours, then it would have taken 75,000 hours to rise 25,000 feet. 75,000 hours is 3,125 days, or 8.56 years.
9. There is a disappointing sidelight to this anecdote. Henry Morris used this argument for the age of the earth against me the first time we debated in April of 1981. I answered it using the information presented here, and he withdrew the argument. It turns out that I was not the first scientist to point out the error in this argument, and you might think, having been shown the mistake, that the creationists would have immediately corrected their literature. Not so. In fact, creationist speakers continue to use the magnetic field as evidence for a young earth, and this long-discredited argument remains in the anti-evolution literature.
10. The data shown here are taken from G. Brent Dalrymple's book, *The Age of*

the Earth (Stanford, CA: Stanford University Press, 1991), p. 377.

11. ^{236}U is a rare nuclide produced by slow neutrons in uranium ore, and ^{129}I is produced by the bombardment of tellurium by cosmic ray muons.
12. I could have made this list *much* longer by including nuclides with half-lives of less than 1 million years. That would have made a table so long that my editors—who have barely tolerated this example—would surely have balked. However, the result would have been the same. The only nuclides found in nature with half-lives of less than a million years are those that are continually produced by natural processes. All the rest are gone, vanished with the passage of time.
13. The actual numbers are 1/1024 and 1/1,048,576.
14. In an effort at being thorough, geologist Brent Dalrymple, who compiled this data, allowed that there were at least two other possibilities for this line in the sand. The first is sheer dumb luck, and the second is some bias in favor of long half-lives by the processes in stars that produce the nuclides of natural elements. The former can be ruled out on statistical grounds, and the latter on experimental evidence showing no such bias.
15. Whitcomb & Morris, *Genesis Flood*, pp. 343–344.
16. Dalrymple, *Age of the Earth*.

    Dalrymple's introduction to this excellent work explains that his encounters with scientific creationists in the 1980s were the driving force behind his decision to write such a text. He was one of several expert witnesses who testified on behalf of evolution at the Arkansas creation science trial in December 1981.
17. Whitcomb & Morris, *Genesis Flood*, p. 344.
18. Ibid, p. 354.
19. Ibid, p. 346.
20. Ibid, p. 369.
21. For a description of this approach, see "The Atomic Constants, Light, and Time," by Norman Trevor and Barry Setterfield, Flinders University of South Australia, School of Mathematical Sciences Technical Report, 1987; also an earlier article by Setterfield, "The Velocity of Light and the Age of the Universe," *Ex Nihilo* 4 (1981): 38.
22. Whitcomb & Morris, *Genesis Flood*, p. 346.

CHAPTER 4

1. The second law of thermodynamics states that in any closed system, energy becomes less available over time. In practical terms, this means that chemical and molecular disorder increases in association with any real process.

Because evolution requires an increase in order and complexity, this means that it runs counter to the second law of thermodynamics, or so the argument goes. However, the second law applies to systems as a whole, not to individual parts of those systems. This means that some parts of a system may indeed become more orderly and complex so long as this increase in order is balanced by an equal or greater decrease in order elsewhere. This, of course, is exactly what living things do—as they grow and evolve they use enormous amounts of energy (usually in the form of food), producing a thermodynamic balance in the system as a whole. The second law no more forbids evolution than it forbids a tiny seed from growing into a larger, more complex tree. Both processes require energy to proceed, and both are in perfect accordance with the laws of thermodynamics.

2. R. Lewin, "Evolutionary Theory under Fire," *Science* 210 (1980): 883–887.
3. I. Kristol, "Room for Darwin and the Bible," *The New York Times* Op-Ed page (Sept. 30, 1986).
4. Darwin, *The Origin*, p. 191.
5. Ibid, p. 203.
6. Ibid, p. 204.
7. S. J. Gould as quoted by R. Lewin in "Evolutionary Theory under Fire," *Science* 210 (1980): 883–887.
8. S. J. Gould and N. Eldredge, "Punctuated Equilibrium Comes of Age," *Nature* 366 (1993): 223–227.
9. From "How Did We Get Here," a 1996 Internet debate between Kenneth Miller and Phillip Johnson, hosted at the *Nova* Web site: http://www.pbs.org/wgbh/nova/odyssey/debate
10. P. E. Johnson, *Darwin on Trial* (Washington, DC: Regnery Gateway, 1991), p. 14.
11. Johnson had used that status earlier in a crusade against what he called the establishment view that human immunopathy virus (HIV) causes AIDS. For several years, he enthusiastically endorsed the arguments of virologist Peter Duesberg that HIV did not cause AIDS, despite mounting scientific evidence to the contrary. Duesberg's efforts against the HIV theory were well cataloged in an article in the December 8, 1997 issue of *The Scientist*. Phillip Johnson had endorsed Duesberg in an earlier issue of the same magazine (January 23, 1995, p. 13).
12. Johnson, *Darwin on Trial,* p. 74. The reference is from B. J. Stahl's textbook, *Vertebrate History: Problems in Evolution* (Mineola, NY: Dover, 1985).
13. Actually, it's generally a lot faster than that. This is a point we will explore below. Studies that have attempted to measure the rate of evolutionary

change produced by natural selection have repeatedly shown that it occurs at rates several orders of magnitude *greater* than the rates observed in the fossil record. Stephen Jay Gould made exactly this point in his column in *Natural History* (Dec. 1997/Jan. 1998, p. 12): "Biologists have documented a veritable glut of cases for rapid and eminently measurable evolution on timescales of years and decades."

14. D. Berlinski, "The Deniable Darwin," *Commentary* 101 (June 1996).
15. Darwin, *The Origin*, p. 462.
16. J. Shoshani, "It's a Nose! It's a Hand! It's an Elephant's Trunk!" *Natural History* (Nov. 1997): 36–45.
17. Data from V. J. Maglio, "Origin and Evolution of the Elephantidae," *Transactions of the American Philosophical Society* 63, no. 3 (1973): 1–149.
18. From Johnson's review, "Daniel Dennett's Dangerous Idea," *New Criterion* (October 1995), 9–14.
19. My source for these figures is a personal communication from the late paleontologist Jack Sepkoski, who has carried out an important series of careful and systematic studies of the fossil record. These include J. Sepkoski, *Paleobiology* 4 (1978): 223–251; J. Sepkoski et al., *Nature* 293 (1981): 435–437; J. Sepkoski, *Paleobiology* 10 (1984): 246–267; and Foote et al., *Science* 283 (1999): 1310–1315.
20. Stephen Stanley, in his book *Macroevolution* (Baltimore: Johns Hopkins University Press, 1998) notes that the horseshoe crab group (the superfamily Limulacea) shows a greater degree of evolutionary change than is often assumed (p. 124). This view is consistent with Stanley's descriptions of the fossil records of sixteen animals widely regarded as living fossils. In each case, the living species are relatively recent appearances in lineages in which the long-term rate of speciation is relatively slow.
21. Johnson, *Darwin on Trial*, p. 98.
22. See Blattner et al., "The Complete Genome of *E. coli*," *Science* 277 (Sept. 1997): 1453–1462.
23. E. C. C. Lin et al., "Evolution of an *Escherichia coli* Protein with Increased Resistance to Oxidative Stress," *Journal of Biological Chemistry* 273 (1998): 8308–8316.
24. Phillip Johnson from the 1996 debate on evolution with Kenneth R. Miller at the *Nova* Web site:
    http://www.pbs.org/wgbh/nova/odyssey/debate
25. This may seem to be an arbitrary figure, but it's not. To correct for problems of scaling, the rate of change is always expressed in logarithmic terms. The number 2.718 is the value of *e,* the number used as the basis

for natural logarithms. The actual formula used to calculate the rate of evolutionary change is

$$r = \frac{\ln x_2 - \ln x_1}{\Delta t}$$

26. D. Reznick et al., "Evaluation of the Rate of Evolution in Natural Populations of Guppies (*Poecilia reticulata*)," *Science* 275 (1997): 1934–1936.
27. Ibid.
28. Gingerich, "Rates of Evolution: Effects of Time and Temporal Scaling," *Science* 222 (1983): 159–161.
29. Darwin, *The Origin*, p. 119–120.
30. S. J. Gould and N. Eldredge, "Punctuated Equilibrium Comes of Age," *Nature* 366 (1993): 223.
31. Ibid, 223.
32. S. J. Gould, *Eight Little Piggies* (New York: W. W. Norton, 1993), p. 277.
33. G. Goodfriend and S. J. Gould, "Paleontology and Chronology of Two Evolutionary Transitions by Hybridization in the Bahamian Land Snail *Cerion*," *Science* 274 (1996): 1894–1897.
34. S. J. Gould, "The Paradox of the Visibly Irrelevant," *Natural History* (Dec. 1997): 64.
35. Johnson has even declined direct invitations to present *any* alternative to evolution. In our Nova Internet debate in 1996 I asked him point-blank to let everyone know his alternate theory. He responded by writing, "I'm not proposing another theory; I'm explaining why I'm not convinced by yours."
36. Johnson, *Darwin on Trial*, p. 75.
37. M. I. Coates and J. A. Clack, "Fish-like Gills and Breathing in the Earliest Known Tetrapod." *Nature* 352 (1991): 234–236.
38. The earliest true vertebrates appear in the early Ordovician, although there is some possibility that the first vertebrates appeared in the very late Cambrian period.

Chapter 5

1. Biology has just begun to approach the point where it, like physics and chemistry, can become a predictive science. This makes it that much more remarkable that Temin successfully predicted the existence of an enzyme for which there was at the time no biochemical evidence. Incidentally, reverse transcriptase is no mere biological curiosity. Not only is this enzyme widely used in research and in biotechnology, it is also used by HIV, the virus that causes AIDS, to insert its genes into an infected cell. Recognizing this, one of the first successful strategies to fight the virus was directed at this very

enzyme. AZT, the well-known anti-AIDS drug is, in fact, an inhibitor of reverse transcriptase.

2. Although Pruisner was awarded the Nobel in 1997, his theories are still controversial, and I'd prefer to wait a few years before placing his prion work in the same category as Temin's prediction of reverse transcriptase.
3. M. J. Behe, *Darwin's Black Box* (New York: The Free Press, 1996).
4. Ibid, p. 5.
5. Ibid, pp. 24–25.
6. Ibid, p. 39.
7. Ibid.
8. From a speech by Michael J. Behe, "Evidence for Intelligent Design from Biochemistry," at the Discovery Institute's God and Culture Conference, Seattle, August 10, 1996.
9. Behe, *Darwin's Black Box,* pp. 232–233.
10. Darwin, *The Origin,* p. 187.
11. Ibid.
12. For a detailed discussion of this remarkable evolutionary transition see A. W. Crompton and F. A. Jenkins, "Origin of Mammals," in *Mesozoic Mammals: The First Two Thirds of Mammalian History*, edited by Z. Kielan-Jaworowska, J. G. Eaton, and T. M. Bown (Berkeley: University of California Press, 1979).
13. Behe, *Darwin's Black Box*, p. 22.
14. Darwin, *The Origin.*
15. From a talk by Michael J. Behe, "Molecular Machines—Experimental Support for the Design Inference." Given in the summer of 1994, the C. S. Lewis Society, Cambridge University.
16. Sperm cells of the genus *Culex* are indeed missing one of the central pair tubules, but they also have an extra circle of microtubules, making their actual arrangement 9+9+1. See Phillips, *Journal of Cell Biology* 40 (1969): 28–43.
17. D. M. Wooley, "Studies on the Eel Sperm Flagellum," *Journal of Cell Science* 110 (1997): 85–94.
18. J. Schrevel and C. Besse, "Un Type Flagellaire Fonctionnel de Base 6+0," *Journal of Cell Biology* 66 (1975): 492–507.
19. Prensier et al., "Motile Flagellum with a '3+0' Ultrastructure," *Science* 207 (1979): 1493–1494.
20. Ibid.
21. Atwell et al., "Structural Plasticity in a Remodeled Protein-Protein Interface," *Science* 278 (1997): 1125–1128.

A little poetic license has been used here. We cannot *watch* proteins doing much of anything in the literal sense, so these researchers employed a laborious, but effective, way of doing much the same thing. They crystallized the proteins and then determined their detailed, atomic-level structure by X-ray crystallography.

22. Hall's experiments describing this system appear in two papers: B. G. Hall, "Evolution on a Petri Dish. The Evolved β=Galactosidase System as a Model for Studying Acquisitive Evolution in the Laboratory," *Evolutionary Biology* 15 (1982): 85–150; and B. G. Hall, "Evolution of New Metabolic Functions in Laboratory Organisms," in *Evolution of Genes and Proteins*, edited by M. Nei and R. K. Koehn (Sunderland, MA: Sinauer Associates, 1983).
23. D. J. Futuyma, *Evolution* (Sunderland, MA: Sinauer Associates, 1986), pp. 477–478.
24. Behe, *Darwin's Black Box,* p. 185.
25. D. J. DeRosier, "The Turn of the Screw: The Bacterial Flagellar Motor," *Cell* 93 (1998): 17–20.
26. A. M. Dean, "The Molecular Anatomy of an Ancient Adaptive Event," *American Scientist* 86 (Jan.-Feb. 1998): 26–37.
27. J. M. Logsdon and W. F. Doolittle, "Origin of Antifreeze Protein Genes: A Cool Tale in Molecular Evolution," *Proceedings of the National Academy of Sciences* 94 (1997): 3485–3487.
28. S. M. Musser and S. I. Chan, "Evolution of the Cytochrome C Oxidase Proton Pump," *Journal of Molecular Evolution* 46 (1998): 508–520.
29. Behe, *Darwin's Black Box,* p. 176.
30. E. Meléndez-Hevia, Waddell, and Cascante, "The Puzzle of the Krebs Citric Acid Cycle: Assembling the Pieces of Chemically Feasible Reactions, and Opportunism in the Design of Metabolic Pathways during Evolution," *Journal of Molecular Evolution* 43 (1996): 293.
31. Ibid, 302.
32. Behe, *Darwin's Black Box,* p. 97.
33. One of the very first human genes to be cloned and sequenced was the Factor VIII gene, with the happy result that large amounts of Factor VIII protein are now produced cheaply and safely by recombinant DNA technology. With regular injections of synthetic Factor VIII, most hemophiliacs can clot their blood normally, and are able to lead healthy lives.
34. The tendency of water to move across membranes such as those that surround cells in the body's tissues creates an *osmotic* pressure strong enough to burst most cells, unless that pressure is offset by the presence of enough soluble material in the blood plasma to balance the soluble material found within the cell. This is why circulatory systems contain substantial amounts of soluble proteins, such as the serum albumin found in human blood.

35. The targeting of proteins to different locations is common and well understood. In a 1996 essay, Neil Smalheiser presented a long list of such cases, noting that mistargeting has "implications for the evolution of cell organization," and could serve "as a source of variation for natural selection." N. R. Smalheiser, "Proteins in Unexpected Locations," *Molecular Biology of the Cell* 7 (1996): 1003–1014.

    More recently, the idea that evolution is driven by the acquisition of additional functions by a variety of cellular proteins was explored by Constance Jeffery of Brandeis University. Jeffery presented a long list of cellular proteins that have acquired additional functions, elegantly documenting the evolution of new biochemical functions within the cell. C. J. Jeffery, "Moonlighting Proteins," *Trends in Biochemical Sciences* 24 (1999): 8–11.

36. Despite his well-informed appreciation of this marvelous system, Behe did two things in his book which I find particularly strange for a biochemist. First, he did not call these factors by their current names. Instead, he used an archaic terminology that has long disappeared from modern textbooks. Factor X is "Stuart Factor," Factor XII is "Hageman Factor," Factor IX is "Christmas Factor," and so on. Second, he described the beautiful cascade of factors as a "Rube Goldberg" machine, comparing it to one of those delightfully intricate contraptions of comically dissimilar parts made famous by the late cartoonist. As should be clear from Figure 5.3, the clotting cascade is not at all like one of Mr. Goldberg's machines, because the key steps of the clotting cascade are *not* dissimilar; in fact, they involve a series of homologous, closely related serine proteases. I cannot say why Behe used an outdated terminology and an inappropriate comparison to make the clotting factors appear dissimilar, but I do wonder if he felt that allowing his readers an understanding of the similarity of the six serine proteases would have weakened his argument that the cascade could not have evolved by gene duplication.

37. See, for example, Lubert Stryer's respected and authoritative textbook, *Biochemistry,* for a description of thrombin's structural and biochemical resemblance to trypsin. L. Stryer, *Biochemistry,* 4th ed. (New York: W. H. Freeman, 1995) p. 254.

38. See, for example, R. F. Doolittle and D. F. Feng, "Reconstructing the Evolution of Vertebrate Blood Coagulation from a Consideration of the Amino Acid Sequences of Clotting Proteins," *Cold Spring Harbor Symposia on Quantitative Biology* 52 (1987): 869–874. A much less formal presentation of the process was given in R. F. Doolittle, "The Evolution of Vertebrate Blood Coagulation: A Case of Yin and Yang," *Thrombosis and Hemostasis* 70 (1993): 24–28.

39. X. Xu and R. F. Doolittle, "Presence of a Vertebrate Fibrinogen-like Sequence in an Echinoderm," *Proceedings of the National Academy of Sciences* 87 (1990): 2097–2101.
40. Incidentally, in most animals the synthesis of vitellogenin is limited to adult females. One might wonder, then, how natural selection could favor this evolutionary scheme. The answer turns out to be surprisingly simple. The initial clotting reaction would appear only in females, making it possible for them to clot their blood more effectively than males. A gene that increases chances of survival only in females is still a useful gene, as modern efforts to cultivate lobsters (which focus almost exclusively on promoting the survival of egg-laying females) show. Once that gene duplication had produced a primitive fibrinogen gene from vitellogenin, a simple mutation in regulatory sequences that switched the gene on in *both* males and females would enhance the survival value of the gene even more. Comparative studies clearly show that this scheme is reasonable. In sea urchins, both sexes produce large amounts of vitellogenin, proving that it is not at all impossible for such a gene to be expressed in both males and females. See R. F. Doolittle and X. Riley, "The Amino-acid Sequence of Lobster Fibrinogen Reveals Common Ancestry with Vitellogenin," *Biochemical and Biophysical Research Communications* 167 (1990): 16–19.
41. Behe, *Darwin's Black Box,* p. 230.
42. Ibid, p. 228.
43. Ibid, p. 249.
44. This is the subtitle of a review, "Darwin v. Intelligent Design (Again)" by H. Allen Orr, which appeared in the December 1996 issue of *The Boston Review*.
45. This debate was held at the 1995 meeting of the American Scientific Affiliation, at Montreat College in North Carolina.

CHAPTER 6

1. David Berlinski wrote, "That Darwin's theory of evolution and biblical accounts of creation play similar roles in the human economy of belief is an irony appreciated by altogether too few biologists." D. Berlinski, "The Deniable Darwin" *Commentary* 101 (June 1996): 19–26.
2. From the Preface of *Teaching About Evolution and the Nature of Science* (Washington, D.C.: National Academy Press, 1998).
3. Ibid.
4. At http://www.nap.edu/readingroom/books/evolution98
5. *Teaching About Evolution and the Nature of Science,* p. 16.

6. Ibid.
7. Figure taken from the National Science Board, Science and Engineering Indicators—1996. (Washington, D.C.: U.S. Government Printing Office).
8. Futuyma, *Evolution,* p. 2.
9. *Teaching About Evolution and the Nature of Science,* p. 58.
10. S. J. Gould, "Nonoverlapping Magisteria," *Natural History* (March 1997): 16–22.
11. P. E. Johnson, "The Unraveling of Scientific Materialism," *First Things* 77 (1997): 22–25.
12. These statements are from a transcript of the program *CBS Sunday Morning* on Nov. 29, 1998. The interviewer was Rita Braver.
13. Dawkins, *Blind Watchmaker,* p. 6.
14. R. Dawkins, *River Out of Eden,* (New York: HarperCollins, 1995), pp. 132–133.
15. W. Provine, "Evolution and the Foundation of Ethics." *MBL Science* 3 (1988): 25–29.
16. W. A. Dembski, ed., *Mere Creation* (Downer's Grove, IL: InterVarsity Press, 1998), p. 14.
17. The teachers vigorously protested the new curriculum to the Tampa Board of Education, and a second vote of the board a few months later canceled the plans to introduce scientific creationism in the Tampa schools. I was happy to have played a small supporting role in their efforts, but the genuine credit for winning this scientific victory in Tampa has to go to the science teachers who organized effectively and energetically to press their case for integrity in science education.
18. J. Hitt, "On Earth as It Is in Heaven," *Harper's Magazine* (Nov. 1996), p. 60.
19. Ford is reported to have alluded to Darwinian natural selection in defense of ruthless capitalism by saying, "This is not an evil tendency in business. It is merely the working-out of a law of nature and a law of God."
20. Quoted in L. S. Feuer, "Is the "Darwin-Marx Correspondence Authentic?" *Annals of Science* 32 (1975): 1–12. This paper analyzes the authenticity of two letters apparently sent by Darwin to Marx. The first, in 1873, acknowledged the gift of *Das Kapital* and thanked Marx, allowing that "I heartily wish I was more worthy to receive it, by understanding more of the deep and important subject of political Economy." The second letter, allegedly written in 1880, appeared only in 1931, and Feuer argues that it had not actually been sent to Marx.
21. My source on this point is a description of the gift in Desmond and Moore's *Darwin: The Life of a Tormented Evolutionist,* p. 601.

22. Dennett wrote, "Let me lay my cards on the table. If I were to give an award for the single best idea that anyone has ever had, I'd give it to Darwin, ahead of Newton and Einstein and everyone else." *Darwin's Dangerous Idea,* p. 21.
23. Ibid, p. 63.
24. I'd like to say "discussion" or "debate," but the tenor of discussion on this issue makes this an appropriate description. Readers interested in following along for themselves might warm up by reading a series of contentious articles and reviews in the *Boston Review*: H. A. Orr, "Dennett's Strange Idea" (Summer 1996); D. Dennett, "The Scope of Natural Selection" (Oct./Nov. 1996); R. C. Berwick, "Feeling for the Organism" (Dec./Jan. 1996–97); R. C. Berwick and J. C. Ahouse, "Darwin on the Mind" (April/May 1998); and S. Pinker, "How the Mind Really Works" (Summer 1998).

    With this background in hand, one can then wade into the peak of the battle as carried on in the *New York Review of Books*: S. J. Gould, "Darwinian Fundamentalism" (June 12, 1997); S. J. Gould, "Evolution: The Pleasures of Pluralism" (June 26, 1997); D. C. Dennett, "'Darwinian Fundamentalism': An Exchange" (Aug. 14, 1997); and finally S. Pinker, "Evolutionary Psychology: An Exchange" (Oct. 9, 1997).

    Andrew Brown has attempted to chronicle these exchanges and more in his book, *The Darwin Wars* (New York: Simon and Schuster, 1999).
25. S. J. Gould and R. Lewontin, "The Spandrels of San Marcos and the Panglossian Paradigm: A Critique of the Adaptationist Programme," *Proceedings of the Royal Society* 205 (1979): 581–598.
26. S. J. Gould, "Tires to Sandals," *Natural History* (April 1989): p. 14.
27. S. Pinker and P. Bloom, "Natural Language and Natural Selection," *Behavioral and Brain Sciences* 13 (1990): 707.
28. Dennett, *Darwin's Dangerous Idea,* p. 391.
29. Ibid, p. 519.
30. Ibid, p. 515.
31. Edward O. Wilson, *On Human Nature* (Cambridge, MA: Harvard University Press, 1978), p. 1.
32. This, in fact, is the theme of a powerful and influential book on behavior, *The Selfish Gene* by Richard Dawkins. If the reader wishes to read a single book presenting a thoroughly modern view of evolution as applied to a variety of biological and social questions, this is unquestionably the one I would recommend. It was also the first, and in my opinion, the very best, of Richard Dawkins's many fine works on evolution.
33. Wilson, *On Human Nature*, p. 188.
34. Ibid, p. 184.

35. Ibid, p. 175.
36. Ibid, p. 167.
37. Dennett, *Darwin's Dangerous Idea*, pp. 517–518.
38. Wilson, *On Human Nature*, p. 192.
39. Although I have used examples of Wilson's reasoning from his 1978 book *On Human Nature*, it is important to note that his views on this subject have not changed. In his article "The Biological Basis of Morality" in the April 1998 *Atlantic Monthly* (p. 65), Wilson wrote, "There is a hereditary selective advantage to membership in a powerful group united by devout belief and purpose. Even when individuals subordinate themselves and risk death in a common cause, their genes are more likely to be transmitted to the next generation than are those of competing groups who lack comparable resolve. . . . Equally important, much if not all of religious behavior could have arisen from evolution by natural selection."
40. D. Hull, "The God of the Galapagos," *Nature 352* (1991): 485–86.
41. From Richard Lewontin's review of Carl Sagan's book, *The Demon-Haunted World: Science as a Candle in the Dark,* in the *New York Review of Books* (January 9, 1997).
42. Ibid.
43. Quoted from "Truly a Wonderful Life," an Internet review by Kurt P. Wise of Stephen J. Gould's *Wonderful Life,* in the *Origins Research Archive* of the *Access Research Network,* vol. 13, no. 1.
44. Bruce Chapman, from the Postscript to Dembski's *Mere Creation*, pp. 457–458.
45. Ibid, p. 458.
46. Lewontin, review of Sagan's *The Demon-Haunted World.*
47. Ibid.

    Lewontin emphasizes this point in a colorful way that reveals his own cultural preconceptions: "The reason that people do not have a correct view of nature is not that they are ignorant of this or that fact about the material world, but that they look to the wrong sources in their attempt to understand. It is not simply, as Sherlock Holmes thought, that the brain is like an empty attic with limited storage capacity, so that the accumulated clutter of false or useless bits of knowledge must be cleared out in a grand intellectual tag sale to make space for more useful objects. It is that most people's mental houses have been furnished according to an appallingly bad model of taste and they need to start consulting the home furnishing supplement of the Sunday *New York Times* in place of the stage set of *The Honeymooners.*"
48. Lewontin, review of Sagan's *The Demon-Haunted World.*

CHAPTER 7

1. Carl Sagan, *The Demon-Haunted World* (New York: Ballantine Books, 1996), p. 30.
2. This quotation was taken from page 163 of *Darwin's Dangerous Idea*, where it is attributed to a translation in Ellegård, 1956, p. 176.
3. My best high school Latin translation of this phrase is "God creates. Linnaeus arranges."
4. Quotation attributed to Heisenberg taken from *Quantum Reality* by Nick Herbert (New York: Anchor, 1997, p. 55).
5. W. Heisenberg, *Physics and Philosophy* (New York: Harper Brothers, 1958).
6. F. Capra, *The Tao of Physics*, 3rd ed. (Boston: Shambhala Publications, 1991).
7. A few of the many books presenting general background on quantum theory intended for the general audience are:

   P. C. Davies, and J. R. Brown (eds.) *The Ghost in the Atom* (Cambridge, England: Cambridge University Press, 1993).

   R. P. Feynman, *Qed: The Strange Theory of Light and Matter*, (Princeton, N.J.: Princeton University Press, 1985).

   J. Gribbin, *In Search of Schrodinger's Cat: Quantum Physics and Reality* (New York: Bantam Books, 1985).

   N. Herbert, *Quantum Reality: Beyond the New Physics* (New York: Anchor, 1997).
8. E. Schrödinger, *What Is Life?* (New York: The University Press, 1945), p. 8.
9. There's an interesting corollary to Schrödinger's assumptions regarding the abilities of our senses to react to individual particles. While his general reasoning on size and the way in which our senses react to stimuli makes good sense, it does indeed look as though under appropriate conditions, our visual system can react to a single photon. This doesn't mean that Schrödinger was wrong. Single-photon detection is possible only when the eyes have been dark-adapted by long periods of total darkness under laboratory-controlled conditions. Under more normal conditions, the eyes, like other senses, average out individual quantum effects to produce the time-averaged sensations of light and darkness that our brains are used to dealing with.
10. Schrödinger, *What Is Life*? p. 68.
11. Ibid, p. 67.
12. Readers will no doubt find other meanings (beyond the literal) in each of these works, but Mann's novel is especially rich in significance. Many critics

have seen Mann's central character as a symbol for the German nation itself, which effectively sold its soul for the Nazi dream of world domination and racial supremacy.

13. S. J. Gould, *Wonderful Life* (New York: W.W. Norton, 1989), p. 51.
14. Ibid, p. 50.
15. For interesting descriptions of this work, see E. Strauss, "How Embryos Shape Up," *Science* 281 (1998): 166–177, or Juan Carlos Izpisúa Belmonte, "How the Body Tells Left from Right," *Scientific American* (June 1999): 46–51.
16. Dembski, *Mere Creation*, p. 14.
17. R. Dawkins, "On Debating Religion," *The Nullifidian* (Dec. 1994).

Chapter 8

1. Carl Sagan quoted in *U.S. News & World Report* (December 23, 1991), p. 61.
2. Actually 7.04 x 10^8 years.
3. S. Hawking, *A Brief History of Time* (New York: Bantam Books, 1988).
4. T. Ferris, *The Whole Shebang* (New York: Simon and Schuster, 1997).
5. R. Jastrow, *God and the Astronomers* (New York: W. W. Norton, 1978), p. 116.
6. Ferris, *The Whole Shebang*, pp. 303–304.
7. Hawking, *A Brief History of Time,* p. 121.
8. B. J. Carr and M. J. Rees, "The Anthropic Principle and the Structure of the Physical World," *Nature* 278 (1979): 605–612.
9. Ibid.
10. F. Dyson, *Disturbing the Universe* (New York: Harper and Row, 1979), p. 250.
11. J. Boslough, *Stephen Hawking's Universe* (New York: William Morrow, 1985), p. 121.
12. Dennett, *Darwin's Dangerous Idea*, p. 166.
13. His actual words are: "If you doubt that the hypothesis of an infinity of variant universes could actually explain this elegance, you should reflect that this has at least as much claim to being a non-question-begging explanation as any traditional alternative." *Darwin's Dangerous Idea*, p. 180.
14. Physicists have long recognized the inability of science, at least at present, to address these questions. Consider this passage from James Trefil, *Moment of Creation* (New York: Collier Books, 1983), p. 223: "But, who created those laws? . . . Who made the laws of logic? . . . No matter how far the

boundaries are pushed back, there will always be room both for religious faith and a religious interpretation of the physical world. For myself, I feel much more comfortable with the concept of a God who is clever enough to devise the laws of physics that make the existence of our marvelous inevitable than I do with the old-fashioned God who had to make it all, laboriously, piece by piece."

15. S. J. Gould in *Wonderful Life* (p. 290) recounts Gray's problems this way: "[Christian evolutionist Asa] Gray, the Harvard botanist, was inclined to support not only Darwin's demonstration of evolution but also his principle of natural selection as its mechanism. But Gray was worried about the implications for Christian faith and the meaning of life. He particularly fretted that Darwin's view left no room for rule by law, and portrayed nature as shaped entirely by blind chance."

Darwin's response, dated May 22, 1860 (from *The Life and Letters of Charles Darwin*), to Gray's hopeful inquiries on the compatibility of evolution and religion illustrates the direction Darwin's own thinking had taken on issues of theology: "With respect to the theological view of the question. This is always painful to me. I am bewildered. I had no intention to write atheistically. But I own that I cannot see as plainly as others do, and as I should wish to do, evidence of design and beneficence on all sides of us. There seems to me too much misery in the world. I cannot persuade myself that a beneficent and omnipotent God would have designedly created the Ichneumonidae [wasps] with the express intention of their [larva] feeding within the living bodies of Caterpillars, or that a cat should play with mice. Not believing this, I see no necessity in the belief that the eye was expressly designed. On the other, I cannot anyhow be contented to view this wonderful universe, and especially the nature of man, and to conclude that everything is the result of brute force. I am inclined to look at everything as resulting from designed laws, with the details, whether good or bad, left to the working out of what we may call chance. Not that this notion at all satisfies me. I feel most deeply that the whole subject is too profound for the human intellect. A dog might as well speculate on the mind of Newton. Let each man hope and believe what he can. Certainly I agree with you that my views are not at all necessarily atheistical. The lightning kills a man, whether a good one or bad one, owing to the excessively complex action of natural laws. A child (who may turn out an idiot) is born by the action of even more complex laws, and I can see no reason why a man, or other animals, may not have been aboriginally produced by other laws, and that all these laws may have been expressly designed by an omniscient Creator, who foresaw every future event and consequence. But the more I think the more bewildered I become; as indeed I probably have shown by this letter. Most deeply do I feel your generous kindness and interest."

16. Gould, *Wonderful Life*, p. 291.
17. Ibid, p. 290.

   As Gould puts it, "Why did mammals evolve among vertebrates? Why did primates take to the trees? Why did the tiny twig that produced Homo sapiens arise and survive in Africa? When we set our focus upon the level of detail that regulates most common questions about the history of life, contingency dominates and the predictability of general form recedes into an irrelevant background."

18. Ian Barbour, physicist and theologian, was awarded the 1999 Templeton Prize for Progress in Religion. He is professor emeritus of physics and religion at Carleton College.
19. I. Barbour, *Religion and Science* (San Francisco: HarperCollins, 1997), p. 216.
20. Barbour, *Religion and Science*, p. 308.
21. See, for example, the chapter entitled "Does God Act in the Physical World," in *Belief in God in an Age of Science* by John Polkinghorne (New Haven, CT: Yale University Press, 1998).
22. J. Polkinghorne, *One World: The Interaction of Science and Theology* (Princeton, NJ: Princeton University Press, 1987), p. 69.
23. S. J. Gould, "On Embryos and Answers," *Natural History* (July/Aug. 1998), p. 58.
24. Calculation: 2 million years / 4.5 billion years. 86,400 seconds in a day x 2/4,500 = 38 seconds.
25. Darwin, *The Origin*, p. 79.
26. The social insects employ a unique system of sex determination called "haplodiploidy," in which all of the worker members of the colony are sterile females. Ordinary sisters are fifty percent related to each other, which means that two systems have a fifty percent chance of sharing any particular gene. However, haplodiploidy ensures that the worker insects are seventy-five percent related to each other—this means that they are *more* than sisters, and dramatically increases the rewards for altruistic behaviors. It is no coincidence that the most well-developed societies (other than ours) in all the animal kingdom are found in those few groups of insects in which haplodiploidy prevails.
27. A. Dillard, *Pilgrim at Tinker Creek* (New York: Harper and Row, 1974), p. 148
28. Barbour, *Religion and Science*, p. 302.
29. From an article by Richard Dawkins ("On Debating Religion") in *The Nullifidian* (Dec. 1994).

30. As quoted in L. C. Eiseley, *Darwin's Century,* (Garden City, N.Y.: Doubleday, 1958), p. 346.
31. C. Iannone, "A Third Way," *First Things* (February 1999), pp. 45–49.
   Exactly this point was made by Carol Iannone in the review of Edward Larson's book on the Scopes trial, *Summer for the Gods*. Iannone describes the anti-evolution stances of David Berlinski, Phillip Johnson, and Michael Behe, but also takes care to note, "None of these thinkers will support as scientifically tenable the literal account of Genesis."
32. H. J. Van Till, "God and Evolution: An Exchange," *First Things* (June/July 1993), pp. 32–41.
33. Augustine, *On the Literal Meaning of Genesis,* Book 1, Chapter 19.
34. Ibid.
   I am particularly grateful to Howard J. Van Till's article in *First Things* for calling my attention to these passages from Saint Augustine, and for providing several excellent arguments which I have done my best to paraphrase here. I particularly recommend Van Till's 1993 article, which is part of a revealing dialogue with Phillip Johnson.
35. This is just one of many apparent contradictions between Genesis 1 and 2. Another famous one is that Genesis 1 apparently describes all plants, including trees, as having been formed on the third day of creation. In Genesis 2, trees are created on the sixth day, following the creation of man but preceeding the creation of woman. Likewise, in Genesis 1, the "waters bring forth fowl," but in Genesis 2, God formed fowl "out of the ground."
36. It won't, incidentally. Dust contains too little carbon and (generally) too much silicon.
37. This point is covered thoroughly in several chapters of Book 4 of *The Literal Meaning of Genesis*, including Chapter 25, "Why no mention is made of night in the six days"; Chapter 26, "The meaning of day in the creation narrative"; Chapter 27, "Days familiar to us are quite different from the days of creation"; and Chapter 29, "Our interpretation does not imply intervals of time in angelic knowledge."

Chapter 9

1. J. Horgan, *The End of Science* (Reading, MA: Addison Wesley, 1996).
2. Ibid, p. 138.
3. Ibid, p. 114.
4. Dean Kenyon's assertion is taken from P. Davis and D. H. Kenyon's *Of Pandas and People* (Richardson, TX: Foundation for Thought and Ethics, 1993). Creationist critics routinely proclaim the absence of transitional fossils, but

rarely are they specific as to where a few transitional forms should be found if evolution were correct. This passage, from pp. 101–102 of *Pandas* is a rare exception: "The absence of unambiguous transitional fossils is illustrated by the fossil record of whales. The earliest forms of whales occur in rocks of Eocine age, dated some 50 million years ago, but little is known of their possible ancestors. By and large, Darwinists believe that whales evolved from a land mammal. The problem is that there are no clear transitional fossils linking land mammals to whales. If whales did have land-dwelling ancestors, it is reasonable to expect to find some transitional fossils. Why? Because the anatomical differences between the two are so great that innumerable in-between stages must have paddled and swam the ancient seas."

Michael Behe, before the publication of *Darwin's Black Box,* used an identical argument in an article included in a volume of essays on the so called evolution-creation issue: "If random evolution is true, there must have been a large number of transitional forms between the *Mesonychid* and the ancient whale. Where are they? It seems like quite a coincidence that of all the intermediate species that must have existed between *Mesonychid* and whale, only species that are very similar to the end species have been found." M. J. Behe, "Experimental Support for Regarding Functional Classes of Proteins to Be Highly Isolated from Each Other" in *Darwinism, Science or Philosophy?* edited by J. Buell and V. Ahearn (Richardson, TX: Foundation for Thought and Ethics, 1994), p. 61.

5. The three species are *Pakicetus inachus*, 52 myr BP; *Ambulocetus natans* 50 myr BP; and *Rodhocetus kasrani* 46–47 myr BP. The discovery of *Ambulocetus natans* was reported in J. G. M. Thewissen, S. T. Hussain, and M. Arif, "Fossil Evidence for the Origin of Aquatic Locomotion in Archaeocete Whales," *Science* 263 (1994): 210–212.
6. Stephen Jay Gould made this point in humorous detail in his column, "Hooking Leviathan by Its Past," in *Natural History* (April 1994), p. 12
7. Futuyma, *Science on Trial,* pp. 12–13.
8. Gould, *Wonderful Life*, p. 318.
9. Ibid.
10. Genesis 1:26.
11. Williams, *Pony Fish's Glow,* pp. 156–157.
12. The work was done at McMaster University in Hamilton, Ontario, by Martin Daly and Margo Wilson. It was described by Jane E. Brody in "Genetic Ties May Be Factor in Violence in Stepfamilies," in *The New York Times* (Feb. 10, 1998).
13. R. Chambers, *Vestiges of the Natural History of Creation* (New York: Humanities Press, republished in 1969), p. 26.

14. Ibid, p. 148.
15. Ibid, p. 389.
16. Ibid, pp. 153–154.
17. Ibid, p. 156.
18. From de Beer's introduction to *Vestiges* in the 1969 edition, p. 31
19. The final edition of *Vestiges,* published in 1884, contained a note from his friend and associate, Alexander Ireland, revealing Chambers as its author.
20. From de Beer's introduction to *Vestiges* in the 1969 edition, p. 33.
21. Darwin was cited by de Beer in his Introduction to *Vestiges,* pp. 33–34.
22. Ibid, p. 34.
23. Wilson, *On Human Nature*, p. 201.
24. S. Pinker, *How the Mind Works* (New York: W.W. Norton, 1997), p. 30.
25. Ibid, p. 555.
26. From "The Evolutionary Psychology of Religion: Does the Brain Have a 'God Module'?" A talk by Steven Pinker on October 14, 1998 at MIT, the "Gods and Computers: Minds, Machines, and Metaphysics" conference.
27. Darwin, *The Origin*, pp. 555–557.
28. Desmond & Moore, *Tormented Evolutionist,* p. 677.
29. Ibid, p. 636. Original quote from his 1879 autobiography, p. 29.
30. F. Darwin, ed., *Life and Letters of Charles Darwin*, vol. 1: (New York: D. Appleton, 1887), p. 282.
31. Darwin, *The Origin,* pp. 550–551.
32. S. J. Gould, *Rocks of Ages*, p. 213.
33. Both quotations from Darwin, *The Origin,* p. 560.

BIBLIOGRAPHY

Appel, T. A. *The Cuvier-Geoffroy Debate.* New York: Oxford University Press, 1987.

Atwell, S., M. Ultsch, A. M. De Vos, and J. A. Wells. "Structural Plasticity in a Remodeled Protein-Protein Interface." *Science* 278 (1997): 1125–1128.

Augustine. *On the Literal Meaning of Genesis* (translated by J. H. Taylor). New York: Newman Press, 1982.

Barbour, I. *Religion and Science.* San Francisco: HarperSanFrancisco, 1997.

Behe, M. J. *Darwin's Black Box.* New York: The Free Press, 1996.

Behe, M. J. "Experimental Support for Regarding Functional Classes of Proteins to be Highly Isolated from Each Other." In *Darwinism, Science or Philisophy?* (Buell, J., and Ahearn, eds.). Richardson, TX: Foundation for Thought and Ethics, 1994.

Belmonte, J. C. I. "How the Body Tells Left from Right." *Scientific American* (June 1999): 46–51.

Berlinski, D. "The Deniable Darwin." *Commentary* 101 (1996): 19–26.

Berwick, R. C. "Feeling for the Organism." *The Boston Review,* (December/ January 1996–97): 23–27.

Berwick, R. C., and J. C. Ahouse. "Darwin on the Mind," Boston Review (April/May 1998): 31–33.

Blattner, F. R., G. Plunkett, C. A. Bloch, N. T. Perna, V. Burland, M. Riley, J. Collado-Vides, J. D. Glasner, C. K. Rode, G. F. Mayhew, J. Gregor, N. W. Davis, H. A. Kirkpatrick, M. A. Goeden, D. J. Rose, B. Mau, and Y. Shao. "The Complete Genome of *E. coli*" *Science* 277 (1997): 1453–1462.

Boslough, J. *Steven Hawking's Universe.* New York: William Morrow, 1985.

Brody, J. E. "Genetic Ties May be Factor in Violence in Stepfamilies." *New York Times,* 10 February, 1998.

Brown, A. *The Darwin Wars.* New York: Simon & Schuster, 1999.

Brown, J. R., P. C. Davies, and J. Broown, (eds.) *The Ghost in the Atom.* Cambridge, England: Cambridge University Press, 1993.

Buffetaut, E., *A Short History of Vertebrate Paleontology.* London: Croom Helm, 1987.

Capra, F. *The Tao of Physics* (3rd ed.). Boston: Shambhala Publications, 1991.

Carr, B. J., and M. J. Rees. "The Anthropic Principle and the Structure of the Physical World." *Nature* 278 (1979): 605–612.

Chambers, R. *Vestiges of the Natural History of Creation.* New York: Humanities Press, 1969. (Originally published 1884)

Chin, K., T. T. Tokaryk, G. M. Erickson, and L. C. Calk. "A King-Sized Theropod Coprolite." *Nature* 393 (1998): 680–682.

Coates, M. I., and J. A. Clack. "Fish-like gills and breathing in the earliest known tetrapod." *Nature* 352 (1991): 234–236.

Crompton, A. W., and F. A. Jenkins. "Origin of Mammals" in *Mesozoic Mammals: The First Two Thirds of Mammalian History,* (Z. Kielan-Jaworowska, J. G. Eaton, and T. M. Bown, eds.). Berkeley: University of California Press, 1979.

Cronin, T. M., and C. E. Schneider. "Climatic Influences on Species: Evidence from the Fossil Record," *Trends in Evolutionary Biology and Ecology* 5 (1990): 275–279.

Daeschler, E. B., and N. Shubin. "Fish with Fingers?" Nature *391* (1997): 133.

Dalrymple, B., *The Age of the Earth.* Stanford, CA: Stanford University Press, 1991.

Darwin, C., *The Origin of Species* (6th ed.). London: Oxford University Press, 1956. (Originally published 1872)

Darwin, F., (ed.). *Life and Letters of Charles Darwin.* New York: D. Appleton, 1887.

Davis, P., D. H. Kenyon. *Of Pandas and People.* Richardson, Texas: Foundation for Thought and Ethics, 1993.

Dawkins, R. "On Debating Religion." *The Nullifidian* (December, 1994).

Dawkins, R. *The Blind Watchmaker.* New York: Norton, 1986.

Dawkins, R. *River Out of Eden.* New York: HarperCollins, 1995.

Dawkins, R. *The Selfish Gene* (new ed.). New York, Oxford University Press, 1989.

Dean, A. M., (1998) "The Molecular Anatomy of an Ancient Adaptive Event." *American Scientist* 86 (January-February 1998): 26–37.

Dembski, W. A., (ed.). *Mere Creation.* Downer's Grove, IL: InterVarsity Press, 1998.

Dennett, D. C., (1997) "'Darwinian Fundamentalism': An Exchange." *New York Review of Books,* 14 August, 1997: 64–65.

Dennett, D. C., (1996) "The Scope of Natural Selection." *Boston Review* (October/November 1996): 34–36.

Dennett, D. *Darwin's Dangerous Idea.* New York: Simon and Schuster, 1995.

DeRosier, D. J. (1998). "The Turn of the Screw: The Bacterial Flagellar Motor." *Cell* 93 (1998): 17–20.

Desmond, A., and J. Moore. *DARWIN: The Life of a Tormented Evolutionist.* New York, Warner Books: 1991.

Dillard, A. *Pilgrim at Tinker Creek.* New York, Harper & Row: 1974.

Doolittle, R. F. "The Evolution of Vertebrate Blood Coagulation: A Case of Yin and Yang." *Thrombosis and Haemostasis* 70 1993: 24–28.

Doolittle, R. F., and D. F. Feng. "Reconstructing the Evolution of Vertebrate Blood Coagulation from a Consideration of the Amino Acid Sequences of Clotting Proteins." *Cold Spring Harbor Symposia on Quantitative Biology* 52 (1987): 869–874.

Doolittle, R. F., and M. Riley. "The Amino-Acid Sequence of Lobster Fibriongen Reveals Common Ancestry with Vitellogenin." *Biochemical and Biophysical Research Communications* 167 (1990): 16–19.

Dyson, F. *Disturbing the Universe.* New York: Harper & Row, 1979.

Strauss, E. "How Embryos Shape Up." *Science* 281 (1998: 166–177.

Ellegård, A. "The Darwinian Theory and the Argument from Design." *Lychnos* (1956): 173–192.

Ferris, T. *The Whole Shebang.* New York: Simon & Schuster, 1997.

Feuer, L. S. "Is the 'Darwin-Marx Correspondence' Authentic?" *Annals of Science* 32 (1975): 1–12.

Feynman, R. P. *Qed: The Strange Theory of Light and Matter.* Princeton, NJ: Princeton University Press, 1985.

Flam, F. (1994) "Co-opting a Blind Watchmaker" Science *265*: 1032–1033.

Foote, M., J. P. Hunter, C. M. Janis, and J. J. Sepkoski. *Science* 283 (1999): 1310–1315.

Futuyma, D. J. *Evolution.* Sunderland, MA: Sinauer Associates, 1986.

Futuyma, D. J. *Science on Trial.* New York: Pantheon, 1983.

Gamow, G. *A Star Called the Sun.* New York: The Viking Press, 1964.

Gingerich, P. D. "Rates of Evolution: Effects of Time and Temporal Scaling." *Science* 222 (1983): 159–161.

Godfrey, L. R. (ed.), *Scientists Confront Creationism*, New York: Norton, 1983.

Goodfriend, G., and S. J. Gould. "Paleontology and Chronology of Two Evolutionary Transitions by Hybridization in the Bahamian Land Snail *Cerion.*" *Science* 274 (1996): 1894–1897.

Gould, S. J. "Nonoverlapping Magisteria." *Natural History* (March 1997): 16–22.

Gould, S. J. "Hooking Leviathan by its Past." *Natural History* (May 1994): 8–15

Gould, S. J. "Darwinian Fundamentalism," *New York Review of Books,* 12 June, 1997: 34–37.

Gould, S. J. "Evolution: The Pleasures of Pluralism." *New York Review of Books,* 26 June, 1997: 47–52.

Gould, S. J. "The Paradox of the Visibly Irrelevant." *Natural History* (December 1997/January 1998): 12–66.

Gould, S. J., and N. Eldredge. "Punctuated Equilibrium Comes of Age." *Nature* 366 (1993): 223–227.

Gould, S. J., and R. Lewontin. "The Spandrels of San Marco and the Panglossian Paradigm: A Critique of the Adaptationist Programme." *Proceedings of the Royal Society* ¨*B* 205 (1979): 581–598.

Gould, S. J. *Eight Little Piggies*. New York: Norton, 1993.

Gould, S. J. *Rocks of Ages*. New York: Ballantine Publishing, 1999.

Gould, S. J. "Tires to Sandals." *Natural History* (April 1989): 8–15.

Gribbin, J. *In Search of Schrodinger's Cat: Quantum Physics and Reality.* New York: Bantam Books, 1985.

Hall, B. G. "Evolution on a Petri Dish. The Evolved β=Galactosidase system as a Model for Studying Acquisitive Evolution in the Laboratory." *Evolutionary Biology* 15 (1982): 85–150.

Hall, B. G. "Evolution of New Metabolic Functions in Laboratory Organisms." in *Evolution of Genes and Proteins*, M. Nei and R. K. Koehn (eds.). Sunderland, MA: Sinauer Associates, 1983.

Hawking, S., *A Brief History of Time*. New York: Bantam Books, 1988.

Heisenberg, W., *Physics and Philosophy*. New York: Harper Brothers, 1958.

Herbert, N. *Quantum Reality: Beyond the New Physics*. New York: Anchor, 1997.

Hitt, J. "On Earth as It Is in Heaven." *Harper's Magazine* (November 1996): 51–60.

Horgan, J. *The End of Science*. Reading, MA: Addison Wesley, 1996.

Hull, D. "The God of the Galapagos." *Nature* 352 (1991): 485–86.

Kristol, I. "Room for Darwin and the Bible." *New York Times*, 30 September, 1986: Op-Ed page.

Iannone, C. "A Third Way." *First Things* (February 1999): 45–49.

Jastrow, R. C. *God and the Astronomers* (second ed.). New York: Norton, 1992.

Jeffery, C. J. "Moonlighting Proteins." *Trends in Biochemical Sciences* 24 (1999): 8–11.

Johnson, P. E. "Daniel Dennett's Dangerous Idea." *The New Criterion* 14 (1995): 9–14.

Johnson, P. E. "The Unraveling of Scientific Materialism." *First Things* 77 (1997): 22–25.

Johnson, P. E., *Darwin on Trial.* Washington, DC: Regnery Gateway, 1991.

Kitcher, P. *Abusing Science.* Cambridge, MA: Harvard University Press, 1997.

Lewin, R. "Evolutionary Theory under Fire." *Science* 210 (1980): 883–887

Lewontin, R. "A Review of *The Demon-Haunted World: Science as a Candle in the Dark,*" *New York Review of Books,* 9 January, 1997: 28–32.

Logsdon, J. M., and W. F. Doolittle. "Origin of Antifreeze Protein Genes: A Cool Tale in Molecular Evolution." *Proceedings of the National Academy of Sciences (USA)* 94 (1997): 3485–3487.

Lu, Z. E. Cabiscol, N. Obradors, J. Tamarit, J. Ros, J. Aguilar, and E. C. C. Lin. "Evolution of an *Escherichia coli* Protein with Increased Resistance to Oxidative Stress." *Journal of Biological Chemistry* 273 (1998): 8308–8316.

Maglio, V. J. "Origin and Evolution of the Elephantidae," *Transactions of the American Philosophical Society,* 63, no. 3(1973): 1–149.

Mayr, E. *The Growth of Biological Thought.* Cambridge, MA: Belknap Press, 1982.

Meléndez-Hevia, E., T. G. Waddell, and M. Cascante. "The Puzzle of the Krebs Citric Acid Cycle: Assembling the Pieces of Chemically Feasible Reactions, and Opportunism in the Design of Metabolic Pathways During Evolution." *Journal of Molecular Evolution* 43 (1996): 293–303.

Milton, J. *Paradise Lost and Paradise Regained.* C. Ricks, (ed.). New York: Penguin Books, 1968.

Montagu, A. (ed.). *Science & Creationism.* New York: Oxford University Press, 1984.

Morris H. M. *The Remarkable Birth of Planet Earth.* San Diego: Creation Life Publishers, 1972.

Morris H. M. *Scientific Creationism.* San Diego: Creation Life Publishers, 1974.

Musser, S. M., and S. I. Chan. "Evolution of the Cytochrome C Oxidase Proton Pump." *Journal of Molecular Evolution* 46 (1998): 508–520.

National Academy of Sciences, *Teaching About Evolution and the Nature of Science.* Washington, D.C.: National Academy Press, 1998.

Orr, H. A. "Dennett's Strange Idea." *Boston Review* (Summer 1996): 28–32.

Orr, H. A. "Darwin v. Intelligent Design (Again)." *The Boston Review* (December 1996): 28–31.

Pinker, S. "Evolutionary Psychology: An Exchange." *New York Review of Books,* 9 October, 1997: 55–57.

Pinker, S. "How the Mind Really Works." *Boston Review* (Summer 1998).

Pinker, S., and P. Bloom. "Natural Language and Natural Selection." *Behavioral and Brain Sciences* 13 (1990): 707–784.

Pinker, S. *How the Mind Works*. New York: Norton, 1997.

Polkinghorne, J. *Belief in God in an Age of Science*. New Haven, CT: Yale University Press, 1998.

Polkinghorne, J. *One World: The Interaction of Science and Theology*. Princeton, NJ: Princeton University Press, 1987.

Provine, W. "Evolution and the Foundation of Ethics." *MBL Science* 3 (1988): 25–29.

Reznick, D. N., F. N. Shaw, F. H. Rodd, and R. G. Shaw. "Evaluation of the Rate of Evolution in Natural Populations of Guppies *(Poecilia reticulata)*." *Science* 275 (1997): 1934–1936.

Roth, A., and R. R. Breaker. "An Amino Acid as a Cofactor for a Catalytic Polynucleotide." *Proceedings of the National Academy of Sciences (USA)* 95 (1998): 6027–6031.

Sagan, C. *The Demon-Haunted World: Science as a Candle in the Dark*. New York: Ballantine Books, 1996.

Schrödinger, E. *What Is Life?* New York: The University Press, 1945.

Sepkoski, J. J. "A Kinetic Model of Phanerozoic Taxonomic Diversity. I. Analysis of Marine Orders." *Paleobiology* 4 (1978): 223–251.

Sepkoski, J. J. "A Kinetic Model of Phanerozoic Taxonomic Diversity. III. Post-Paleozoic Families and Mass Extinctions." *Paleobiology* 10 (1984): 246–267

Sepkoski, J. J., R. K. Bambach, D. M. Raup, and J. W. Valentine. *Nature* 293 (1981): 435–437.

Setterfield, B. "The Velocity of Light and the Age of the Universe." *Ex Nihilo* 4 (1981): 38.

Shoshani, J. "It's a Nose! It's a Hand! It's an Elephant's Trunk!" *Natural History* (November 1997): 36–45.

Smalheiser, N. R. "Proteins in Unexpected Locations." *Molecular Biology of the Cell* 7 (1996): 1003–1014.

Stahl, B. J. *Vertebrate History: Problems in Evolution*. New York: Dover, 1985.

Stanley, S., *Macroevolution*. Baltimore, MD: Johns Hopkins University Press, 1998.

Stemmer, W. P. C. "Rapid Evolution of a Protein In Vitro by DNA Shuffling." *Nature* 340 (1994): 389–391.

Strahler, A. N. *Science and Earth History.* Buffalo, NY: Prometheus Books, 1987.

Stryer, L. *Biochemistry* (4th ed.). New York: Freeman, 1995.

Thewissen, J. G. M., S. T. Hussain, and M. Arif. "Fossil Evidence for the Origin of Aquatic Locomotion in Archaeocete Whales." *Science* 263 (1994): 210–212.

Trefil, J. *The Moment of Creation*. New York: Collier Books, 1983.

Van Till, H. J. "God and Evolution: An Exchange." *First Things* (June/July 1993): 32–41.

Whitaker, A. *Einstein, Bohr and the Quantum Dilemma*. Cambridge, England: Cambridge University Press, 1996.

Whitcomb, J. C., and H. M. Morris. *Genesis Flood*. Phillipsburg, NJ: Presbyterian and Reformed Publishing, 1961.

Williams, G. C. *The Pony Fish's Glow*. New York: HarperCollins, 1997.

Wilson, E. O. *On Human Nature*. Cambridge, MA: Harvard University Press, 1978.

Wise, K. P. "Truly a Wonderful Life." *Origins Research Archive* 13, no. 1 (1997).

Available at: http://www.leaderu.com/orgs/arn/orpages/or131/wise.htmel

Wooley, D. M. "Studies on the Eel Sperm Flagellum." *Journal of Cell Science* 110 (1997): 85–94.

Xu, X., and R. F. Doolittle. "Presence of a Vertebrate Fibrinogen-like Sequence in an Echinoderm." *Proceedings of the National Academy of Sciences (USA)* 87 (1990): 2097–2101.

INDEX

Page numbers of charts and illustrations appear in italics

Insights,
Interviews
& More . . .

Meet Kenneth R. Miller

Stew Milne

KENNETH R. MILLER grew up in Rahway, New Jersey, and attended the local public schools. He served as president of his high school student council, and in 1965 was elected as the New Jersey Governor of American Legion Boys' State. He is an Eagle Scout, and was a varsity swimmer in both high school and college. Miller attended Brown University, where he majored in biology, and earned his Ph.D. in biology at the University of Colorado in 1974. He spent six years as a lecturer and assistant professor at Harvard, returning to Brown as a faculty member in 1980, and is now a professor in the Department of Molecular Biology, Cell Biology, and Biochemistry. His laboratory research focuses on the structure and composition of cellular membranes, and he has published more than fifty scientific papers in journals such as *Cell*, *The Journal of Cell Biology*, *Nature*, and *Scientific American*.

Miller married his college sweetheart, Jody Zanot, in 1972. They live on a small farm in Rehoboth, Massachusetts, and have two daughters. One is a wildlife biologist and the other a history teacher. From his years as a youth sports coach for his daughters, Ken acquired a love of fast-pitch softball, and has umpired both high school and NCAA softball in New England for more than a decade. He is an avid fan of the Boston Red Sox, and especially enjoys reading the Spenser detective novels of Robert B. Parker.

His reasons for writing *Finding Darwin's God* relate to his interests in science, education, and spirituality, and he describes them this way:

> I went to graduate school fully expecting to pursue a career dedicated solely to scientific research, and imagined I would spend most of that career working at the laboratory bench. On the way, and almost by accident, I discovered the joys of teaching. Working as a tutor and a teaching assistant, I found that I enjoyed explaining complicated ideas, and relished the moment when I could see the light go on in a student's eyes. At Harvard I established a successful lab and won my first research grants, critical milestones in the life of any academic scientist, but I also jumped at the chance to put myself in front of undergraduate classes. When I returned to Brown I quickly volunteered to teach introductory biology courses, and have done so ever since.
>
> Two critical events happened during my first couple of years at Brown that drew me into our country's struggles over evolution.
>
> The first was a public debate in 1981 with Henry Morris, founder of the Institute for Creation Research, and at the time the nation's leading antievolutionist. I was persuaded to debate Morris by a group of my students, and accepted despite the fact that, being a cell biologist, evolution was not really my field. My preparation for ▶

"From his years as a youth sports coach for his daughters, Ken acquired a love of fast-pitch softball, and has umpired both high school and NCAA softball in New England for more than a decade."

that debate, however, convinced me that the scientific distortions and misrepresentations put forward by the antievolution movement presented a profound and serious threat to science that simply had to be countered. Science could not cede the public square to the opponents of science itself. I also discovered that my work in molecular biology and biochemistry put me in an ideal position to counter the arguments often put forward by today's creationists, and set about doing exactly that.

The second was a phone call from a former student and young Boston College professor, Joseph S. Levine, who proposed that we work together on a project that at the time I thought ridiculous: Joe wanted me to help him write a high school biology textbook. It took Joe quite a while to persuade me that this was even possible, but persuade me he did. The results of Joe's smooth talking are a series of high school textbooks that he and I have authored during more than two decades of work together. Millions of high school students in every state have used our textbooks, and this has drawn us directly into conflicts with those who would rewrite science curricula and textbooks to de-emphasize or undermine the teaching of evolution.

"Millions of high school students in every state have used our textbooks, and this has drawn us directly into conflicts with those who would rewrite science curricula and textbooks to de-emphasize or undermine the teaching of evolution."

For many years I described my willingness to engage in debate and discussion over evolution as just an academic "hobby," expecting opposition to evolution to wither away once the pseudo-scientific arguments of the creationists had been addressed and answered. More recently, however, antievolutionism has risen again in the United States, and has proved to be a powerful social and political force, and I began to wonder why. As I spoke at various places around the country, I found that people would often comment that they found the scientific evidence for evolution to be overwhelming, but they still could not accept it. The reason, as I was told over and over again in many ways, was the question of God. Evolution simply was not compatible with a loving and purposeful Creator, and that was that. The depth and breadth of this sentiment shocked me, especially since I knew that traditional Christian theology had long since accepted the validity of natural processes as instruments of God's plan and purpose. I found myself telling my audiences that their misgivings about evolution had been answered by Augustine and Aquinas centuries before Darwin, and pointing out that faith and reason are both gifts from God. To accept only faith and to reject scientific reason would be a profound misuse of our human legacies. Time and time again, they asked me to explain. And, time and time again, I did.

These questions ultimately became my reason for writing *Finding Darwin's God.* It had become apparent to me that most Americans regarded a purely scientific defense of evolution as inadequate unless one also answered the question, "What about God?" As a person of faith, I thought I knew that answer, and I wanted to find a way to share it with others. Ultimately, my greatest satisfaction as an author has come from the personal messages of support and appreciation from so many who share ▶

my view that science and religion engage each other in a way that enriches both and completes our understanding of what it means to be alive in this remarkable universe. ❧

Darwin's Day in Court
Dover and the Trial of "Intelligent Design"

About the book

LIKE MOST AMERICANS, I have long regarded the Scopes "Monkey Trial" of 1925, as a landmark event in our nation's history. Despite the fact that biology teacher John Scopes was convicted of violating Tennessee's antievolution law, the trial exposed the lack of scientific evidence for any alternative to evolution, and led to public mockery of the creationist point of view as championed by William Jennings Bryan. But Scopes settled nothing. Laws against the teaching of evolution remained on the books until the 1960s, and opposition to Darwin has held steady despite their repeal. In the 1980s, Arkansas and Louisiana passed laws mandating the teaching of "creation science" alongside evolution, and other states contemplated doing the same. Even today, some states paste warning stickers inside their students' textbooks, admonishing them to remember that evolution is just "a theory, not a fact."

In 1987, the creation science movement was dealt an apparently fatal blow in the case of *Edwards v. Aguillard*. By a definitive 7–2 vote, the U.S. Supreme Court ruled that creationism and creation science were inherently religious; therefore, no public school system could teach them without becoming an advocate of a particular religious point of view. Such advocacy, the Court observed, is prohibited by the First Amendment to the U.S. Constitution. Case closed, or so one might have thought.

In truth, the creation science movement ▸

"[The Scopes 'Monkey Trial'] settled nothing. Laws against the teaching of evolution remained on the books until the 1960s, and opposition to Darwin has held steady despite their repeal."

Darwin's Day in Court *(continued)*

did pause for a while, searching for a new strategy that might find its way around the Court's ruling. The name of that new strategy was "intelligent design." Its intellectual godfather was Phillip Johnson, a Berkeley law professor.

Courtesy of Art Lien (courtartist.blogspot.com)

Johnson worked tirelessly to recruit the opponents of evolution to this new movement, urging them to abandon their differences on matters such as the age of the earth and the nature of the fossil record, and focus instead on just one issue: the need for a "designer" to account for the complexity of life. In short order, the new movement gained an inner circle of advocates with professional credentials; a wealthy sponsor, the Discovery Institute of Seattle; and even a textbook, *Of Pandas and People*, published by the Texas Foundation for Thought and Ethics. Books and pamphlets authored by this network sought to make "intelligent design" (ID) intellectually respectable, and urged educators to include it in their schools.

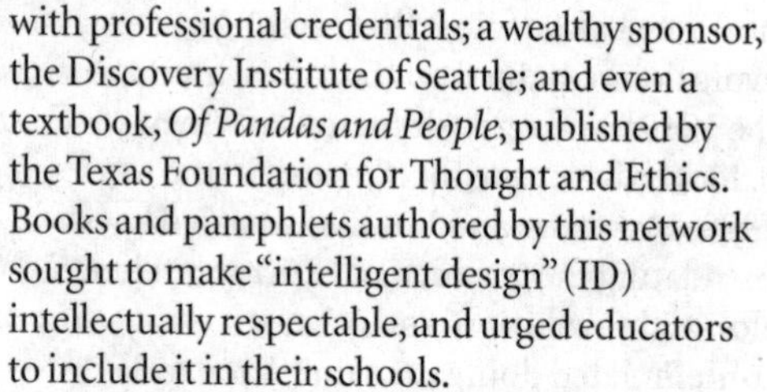

Although several states, including Ohio and Kansas, toyed with including ID in their schools, only one school board actually took the plunge. In October 2004, the School Board of Dover, Pennsylvania, voted to add ID to their high school biology curriculum. The Board placed classroom sets of the *Of Pandas and People* text in the school library, and drafted a four paragraph statement on ID to be read to students. When Dover's biology teachers refused to read the statement, the Board sent its professional employees into the classroom in their places. Shortly thereafter, eleven parents filed a First Amendment lawsuit against the Board in the Pennsylvania Federal Court. *Kitzmiller v. Dover*, as that case is known, went to trial in September 2005.

Early in 2005 I was asked to serve as an expert witness for the parents, and I gladly agreed. In a sense I was already connected with the case, having coauthored (with Joe Levine) the biology textbook used in Dover High. Other expert witnesses lined up on both sides of the case, including a string of prominent ID advocates from the Discovery Institute, and it seemed that "Scopes II" (or maybe "Scopes III") was about to begin. Lawyers defending the School Board made it clear that they wanted to make the case that ID was a legitimate scientific theory, and therefore its inclusion in the school curriculum would serve a clear secular purpose. Judge John E. Jones, presiding over the case, indicated that he would give the Board as much time and as many witnesses as they wanted to make that case.

Then, in mid-summer of 2005, as I and the other expert witnesses for the plaintiffs prepared our trial exhibits and underwent day-long depositions by lawyers for the Board, the darnedest thing happened. The Discovery Institute's Center for Science and Culture, the national champions of ID, turned and ran, in what they claimed was a dispute over personal legal counsel for the case. Five of the eight expert witnesses they had promised to help the Board, including William Dembski, Stephen Meyer, and Warren Nord, withdrew from the case. Perhaps they saw what was coming—a train wreck for the forces of ID.

I spent the first two days of the trial on the witness stand, making the scientific case for evolution and anticipating the arguments that might be made by the ID witnesses that would follow, notably Michael Behe of Lehigh University. Behe's arguments, which I consider briefly in Chapter 5 of *Finding Darwin's God*, center around the claim ▶

"Early in 2005 I was asked to serve as an expert witness for the parents, and I gladly agreed. In a sense I was already connected with the case, having coauthored the biology textbook used in Dover High."

that cells contain complex biochemical machines that are "irreducibly complex," in that they require all of their component parts in order to function. Since, as Behe has pointed out, natural selection can only select for machines that are already working, evolution could never have produced devices like the bacterial flagellum . . . and, if natural selection couldn't do it, then those things must have been "designed." They must be the product of direct supernatural creation.

That argument sounds mighty powerful—until one begins to search through the scientific literature. As I pointed out in my time on the stand, even Behe's favorite example, the bacterial flagellum, actually is composed of smaller submachines with distinct, selectable functions of their own. As a result, the flagellum isn't irreducibly complex at all, and therefore it's not a legitimate argument against evolution. Incredibly, the judge got it. In fact, he got it all.

Judge Jones took notes like a desperate grad student throughout the trial, which *New Yorker* writer Margaret Talbot described as "the biology class you wish you could have taken." He followed intently as I showed the exact point where primate chromosomes 12 and 13 had fused in one of our ancestors to produce human chromosome number 2. He noted the glaring errors of molecular biology in *Of Pandas and People*, and laughed as Berkeley paleontologist Kevin Padian pointed out the incredible misrepresentations of fossils in the same book. His eyes opened wide as Professor Barbara Forrest from Southeastern Louisiana University showed how *Of Pandas and People* had been fashioned from an earlier, overtly creationist book. The publishers had simply run the text through a word processor that

" Judge Jones took notes like a desperate grad student throughout the trial, which *New Yorker* writer Margaret Talbot described as 'the biology class you wish you could have taken.' "

changed each occurrence of "Creator" to "designer," and "creation" to "intelligent design." Judge Jones would, in his decision, call these clumsy attempts to conceal the creationist character of *Of Pandas and People* "astonishing," and would note that the definition of "intelligent design" given in the book was, in fact, identical to "the definition for creation science in early drafts." So much for the argument that ID is somehow different from creation science.

On December 21, 2005, Judge Jones released a sweeping decision that demolished the scientific and political pretensions of ID. The advocates of "design" had been given every chance to establish their views as science, and they had failed, even in front of a Republican judge appointed to the Federal bench by George W. Bush. The reasons for that failure were obvious to anyone who attended the trial, even to those who heard only the case for the School Board. Indeed, in many respects the Board's expert witnesses for ID had destroyed their own case. Each admitted that ID was inherently religious, each acknowledged that there are no peer-reviewed research papers supporting either irreducible complexity or intelligent design, and each failed to refute even a single piece of the scientific evidence assembled by the plaintiffs, the parents who filed the lawsuit to overturn the Board's ID policy. The Dover School Board, in the words of the decision, had displayed "breathtaking inanity" in its willingness to embrace and implement the ID policy. The voters of the Town of Dover, evidently, came to the same conclusion. Six weeks *before* the decision was announced, they voted out of office every member of the pro-ID Board whose name appeared on the November ballot. ▶

"The advocates of 'design' had been given every chance to establish their views as science, and they had failed, even in front of a Republican judge appointed to the Federal bench by George W. Bush."

In the months since the trial and the decision, I've had the occasion to get together several times with my fellow witnesses, our attorneys, the parents who filed the lawsuit, and the Dover teachers. We've gradually come to appreciate just how remarkable an experience the trial was, and how thoroughly all of us felt bonded together in the cause of science. We have begun to sense that the Dover case really is about to become part of history, just like the Scopes trial.

Both trials exposed the great and unfortunate divide between the secular and religious cultures in our country, both involved the issues of what is to be taught in public schools, and both demonstrated the great unease with which the theory of evolution continues to be regarded by Americans. To be sure, there were important differences. The Scopes trial was local, while the Dover case was heard in Federal Court. The point at issue in Dover was not a state law forbidding the teaching of evolution, but a school board policy mandating the teaching of intelligent design. The judge in the Scopes trial did not allow scientific evidence to be considered in his courtroom, while the Dover judge did. And there was one other difference, as well: we won the Dover case. That alone gives me hope that the ultimate battle for the hearts and minds of the American people can be won for the cause of a scientific rationalism that respects both faith and reason.

"The judge in the Scopes trial did not allow scientific evidence to be considered in his courtroom, while the Dover judge did."

An Excerpt from *Only a Theory: Evolution and the Battle for America's Soul*

Only a Theory *will be published in 2008 by Viking Press.*

Chapter 1
Only a Theory

IN A COURTROOM, even a whisper can catch your attention, especially one that comes right at you with a smile and a wink.

"Only a theory" she said, shaking her head just enough to get my attention as I walked past her. "It's only a theory—and we're gonna win." Her smile was genuine and its certainty was unmistakable.

She didn't win—at least not that day and not in that court, but the quiet certainty of the remark has stayed with me ever since. It's provoked me to doubt, wonder, and even fear—and it's my reason for writing this book. It came at the conclusion of my first trial, the first time I have ever sat in a witness stand and given testimony, the first time I'd ever been cross-examined, the first time I'd ever had to meet reporters on the courthouse steps. But it wouldn't be the last—and that's part of the story, too.

When I walked into a Federal courtroom in Atlanta in the fall of 2004, you would have thought that a book was on trial. An attorney stood next to a four-foot-high enlargement of the Table of Contents of a biology textbook. Nearby, a collage of more than fifty pages from the evolution section of the same book had been pasted against cardboard and placed ▸

"When I walked into a Federal courtroom in Atlanta in the fall of 2004, you would have thought that a book was on trial."

An Excerpt from *Only a Theory*
(continued)

on a mounting stand. It seemed to form a wall of evidence that might be used, one could suppose, to convict the book or its authors of some awful, seditious offense against the state or against the good school children of Georgia for whom the book had been written.

These were first impressions, to be sure, but they were first impressions that mattered, especially to me, the coauthor of that book. It was almost as though the project on which Joe Levine and I had labored for so many years had been cut apart, and now its entrails were glued to that board like the organs of a laboratory animal pinned against the soft wax of a dissecting tray. When I climbed into the witness stand, I wondered if the attorneys regarded me that way, too. Were they looking at me as I might look at a laboratory mouse? Trying to find a quick and easy way to get inside and take what they needed?

I would find out soon enough. But the remarkable thing about that trial—the packed courtroom, the media attention, the calls from reporters—was the size and scale of what was being litigated. All of the effort, attention, paperwork, and argument was focused on a paper sticker, about six inches square, containing just three sentences:

> This textbook contains material on evolution. Evolution is a theory, not a fact, regarding the origin of living things. This material should be approached with an open mind, studied carefully, and critically considered.

Why such a fuss? The issue in this trial was whether the actions of the Cobb County Board of Education, in affixing this sticker

“The remarkable thing about that trial was the size and scale of what was being litigated. All of the effort, attention, paperwork, and argument was focused on a paper sticker, about six inches square.”

to the inside of thousands of public school science textbooks, amounted to a violation of the First Amendment of the U.S. Constitution. For the record, the Court found that the stickers were indeed such a violation, and ordered them removed. Constitutional questions are always matters of public interest, and one that applies to the public schools, where most American children are educated, naturally draw plenty of attention and passion.

But it didn't take a psychologist to sense that there was something more at work here, something far deeper than the establishment clause or the narrow scientific meaning of a word like "theory." The stickers were actually the result of a school board's effort to fashion a compromise between thousands of its constituents and the science education standards their schools were required to meet. Those standards had pushed evolution, the central organizing principle of the biological sciences, into the textbooks, classrooms, and even into the homes of families in Cobb County, and thousands of them had pushed back.

They fought back with sermons from pulpits around the county, with protests at School Board meetings, and a massive petition drive demanding that the board either remove evolution from the classroom, or teach it with an alternative more friendly to their views of God's creation. You might say that the sticker was the Board's half-baked effort to split the difference between God and science—and that it satisfied neither.

I'll have more to say about the wording of that sticker a bit later, but what most impressed me at the trial was the passion of ▸

“You might say that the sticker was the Board's half-baked effort to split the difference between God and science—and that it satisfied neither.”

An Excerpt from *Only a Theory*

(continued)

those who defended it. It was clear that something much greater was at stake than a handful of words pasted inside a book, something that would inspire a civic movement—not just here in Georgia, but all across the country. Something that would lead a majority of Americans to say they reject an idea at the very heart of biology, the theory of evolution. To them, what was at issue was a question of the heart and soul. An issue for which they were prepared to fight, and fight they would.